THE GREENPEACE STORY

Michael Brown & John May

"The optimism of the action is better
than the pessimism of the thought"
Harald Zindler

DORLING KINDERSLEY
LONDON

The Authors

One of America's leading environmental journalists, **Michael Brown** is author of the controversial *Laying Waste* (Pantheon), the book that exposed the scandal of toxic waste dumps at Love Canal in the USA, and *The Toxic Cloud* (Harper & Row 1987). He has written extensively for major newspapers and magazines in the USA, including *The New York Times Sunday Magazine*, *Atlantic Monthly*, *Reader's Digest*, *Discover*, and *New York Magazine*.

John May is editorial director of Greenpeace Books and author of *The Greenpeace Book of Antarctica* (1988). The co-author of a number of books including *Weird and Wonderful Wildlife* and *The Book of Beasts*, he has also acted as a consultant to many environmental publications and projects and, as a freelance journalist, has worked for *The Sunday Times*, the *Observer*, the *Guardian*, *BBC Wildlife*, *Cosmopolitan*, *Time Out*, and *The Face*.

GREENPEACE BOOKS
Editorial Director *John May*
Editorial Production and Research *Ian Whitelaw*
Editorial and Research Assistant *Tanya Seton*
Researcher *Raewyn McKenzie*

DORLING KINDERSLEY
Editor *Lesley Riley*
Art Editor *Arthur Brown*
Designers *Jane Warring, Caroline Mulvin*
Computer page make-up *Peter Cooling*

Editorial Director *Jackie Douglas*
Managing Art Editor *Alex Arthur*

Cover Artwork/photomontage by *Shoot That Tiger*
(Whale photograph by *Al Giddings/Ocean Images, Inc.*)

First published in Great Britain in 1989 by
Dorling Kindersley Limited, London

British Library Cataloguing in Publication Data

Brown, Michael
    The Greenpeace story.
    1. Ecology action groups: Greenpeace
    I. Title   II. May, John, *1950-*
    574.5'06'073

    ISBN 0-86318-328-X

Typeset by *The Setting Studio*, Newcastle-upon-Tyne
Reproduced by *Colourscan*, Singapore
Printed and bound in Italy by *Graphicom*

CONTENTS

INTRODUCTION

T HIS BOOK is the most comprehensive account yet written of a remarkable organization that has grown from a handful of determined people into an international network with a membership of more than three million.

Greenpeace began by hiring one battered boat and now owns a sophisticated fleet of ocean-going vessels and river craft. It began by opposing one nuclear test and has now expanded its campaign coverage to include a range of issues such as toxic waste, acid rain, kangaroo slaughter, nuclear weapons at sea, whaling, ocean pollution, and many others as the threats to the natural environment have proliferated. Now firmly established in the Western world, it is busy setting up bases in Latin America and moving into the Soviet bloc. It even has a small research station in Antarctica.

The last fifteen years have seen the transformation not only of Greenpeace but also of the world in which it operates. Ecological issues, which were once on the fringe, have now become the central questions of our time, ranking high on both national and international agendas, and a major topic of public concern.

The Greenpeace story is not a neat and orderly one. It sprawls, loops, twists and expands in every direction as if in defiance of any attempt to contain or define it. This is history written on the run. News was being made as we compiled the book. The tale told here is not the end of the story. Our understanding of the historical forces at work will deepen with time and new chapters will continually have to be added as fresh events crowd in on the old.

In this book we have aimed simply to present a readable account of Greenpeace's most public face, the one for which it is most well known - direct action. It is this approach that has consistently marked it out from other environmental groups and the one that has gained it press and television headlines around the world.

Direct action remains the central theme of Greenpeace operations. This needs to be stated clearly because there is a current media cliché that Greenpeace is turning its back on such tactics and is becoming a more bureaucratic, softer version of its earlier radical self. This is demonstrably untrue; the number of direct actions continues in an upwards spiral. What is true is that, in recent years, such actions have been backed up by sophisticated political lobbying and scientific enquiry that have added strength to the organization's dramatic calls for change. Greenpeace's continued insistence on non-violent tactics, even when faced with violence, reflects both its cultural origins and its links with the other great movements for social change in the twentieth century.

The bombing of the *Rainbow Warrior* transformed the organization, making it headline news around the world and reminding everyone of the forces that are arrayed against it. Consistently working in the face of danger, Greenpeace holds the thin green line. It seeks to transform, radically, both our understanding of the world and the direction in which it is heading. Its message is simple and powerful: everyone has the right to clean water, fresh air and a safe future.

Greenpeace encourages us to see the world as an indivisible whole, to cherish life on Earth, to recognize that national boundaries are false divisions on a natural landscape, to stand up and say enough is enough. By placing itself between the natural world and the forces that seek to destroy it, Greenpeace is acting for us all.

Like any human organization Greenpeace has its faults and weaknesses. Yet the sum of its actions, carried out for the greater good, often with selfless disregard for personal safety, is impressive. Greenpeace has already made a major contribution to limiting nuclear tests, saving wildlife and alerting the world to the damage being done to the fabric of the planet.

Greenpeace is a force for good and also a force for hope, the hope that we can find a solution to the environmental problems that beset us.
By taking action it reminds us that change is possible.
It is also essential.

CHAPTER ONE

DON'T MAKE A WAVE

SET IN THE ICY WATERS off the west coast of Alaska, near the tip of the long, scattered arc of the Aleutians, the tiny island of Amchitka was a haven for wildlife, home to bald eagles and peregrine falcons, and the last refuge of 3,000 endangered sea otters. Signs on the rocky shore advised visitors that it was unlawful to discharge firearms there.

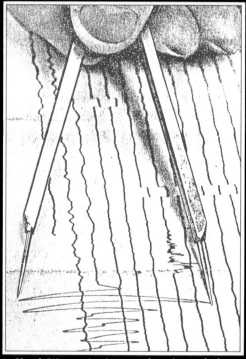

Shock Waves *A seismograph registers the huge force of a US nuclear blast at Amchitka.*

Amchitka is situated in one of the most earthquake-prone regions in the world. In 1964 a massive quake, registering between 8.3 and 8.6 on the Richter scale, had cut an 800-kilometre- (500-mile-) wide swathe of destruction across Alaska, killing 115 people, making thousands homeless and destroying 75 per cent of the state's commerce and industry. It generated a series of tsunamis, seismic sea-waves, that crashed onto the beaches of Oregon, California, Hawaii and Japan. Over the next 18 months there were 10,000 after-shocks.

This was the unlikely site that the United States of America chose for the testing of its nuclear arsenal. On October 2, 1969, Amchitka was rocked by the force of a one-megaton nuclear bomb exploding 1,200 metres (4,000 feet) beneath its surface.

Code-named Milrow, the blast had been surrounded by controversy, primarily because people were worried that it would trigger another disastrous earthquake. On the day of the test, 10,000 protesters blocked the major US-Canadian border crossings to demonstrate their deep concern. Their banners read: "Don't Make A Wave. It's Your Fault If Our Fault Goes". It was the first time the border between the two countries had been closed since the War of 1812, when US troops had fought against the British in Canada.

The US Atomic Energy Commission (AEC) had ignored all protests and proceeded with the test at 3.06 that Thursday afternoon. In Victoria, Canada, seismographs recorded shock waves measuring 6.9 on the Richter scale. But the bomb caused neither an earthquake nor a tidal wave, and people's fears were allayed for a while. Then came the US announcement of plans for a 1971 test blast – code-named Cannikin – five times stronger than Milrow. Something had to be done.

JOINING FORCES

One of those most concerned with the anti-nuclear protest was Jim Bohlen. A deep-sea diver and radar operator in the US Navy during World War II, New Yorker Bohlen had been in the Pacific during the Okinawa and Iwo Jima campaigns; he was at Okinawa when the USA dropped the first atomic bombs on Hiroshima and Nagasaki in 1945.

After the war, Bohlen's skills led him to work on Minuteman and Polaris missiles – key items in the USA's growing arsenal of space-age weaponry. But Bohlen had been disturbed by the threat of nuclear war posed by the Cuban Missile Crisis in 1962, and he became increasingly alienated from his government's policies. He strongly objected to the US involvement in Vietnam and when his stepson became eligible for the draft in 1966, Bohlen resigned his job and left with his family for Vancouver, where he worked as a forest-products research scientist.

Once there, Bohlen and his wife Marie were soon actively involved with the peace movement, helping other draft dodgers by putting them up in their home. With the activism came important new acquaintances. During an anti-war march in 1967, the Bohlens met a Quaker couple, Irving and Dorothy Stowe.

The Stowes had also fled the United States, leaving for New Zealand in 1961. "We thought there might be a chance of survival if there was a holocaust in the northern hemisphere, and… we really didn't want our

> " *The Monday noon demonstration against the Amchitka Island A-bomb test has begun.*
>
> *There are maybe a hundred of us at the beginning, somewhere between 200 and 300 near the end, an hour later.*
>
> *Who are we? A collection, initially, of very proper and respectable and decently paid and serious and a bit less than illiterate citizens, some professors and some ministers and housewives involved in the Society for Pollution and Environmental Control who have just recently discovered that, as one speaker will put it shortly, 'there is no doubt any longer that if things continue the way they are for another 20 years we will all be dead.'*
>
> *And so there are women who are starting to learn the tricks of organization. That gives me more hope than anything has in years. The power of aroused mothers is famous.*
>
> *Politicians, take note. There is a power out there in suburbia, so far harnessed only to charity drives, campaigns and PTAs which, if ever properly brought to bear on the great problems of the day, will have an impact so great the result of its being detonated (like the Amchitka A-bomb test) cannot be predicted.* "
>
> Robert Hunter, Vancouver *Sun*, October 2, 1969

children subjected to the terror of bomb drills," recalls Dorothy. In 1966, they moved to Vancouver, where Irving, a Yale-educated lawyer from New England, soon found an outlet for his strong anti-war sentiments in the *Georgia Straight*, an underground newspaper.

It was Irving who introduced Bohlen to the Quaker religion. Quakers believe in a form of protest known as "bearing witness" – a sort of passive resistance that involves going to the scene of an objectionable activity and registering opposition to it simply by one's presence there.

Bohlen and Stowe had also become active members of the Canadian branch of the Sierra Club, an American conservation group, and in 1969, after fighting a number of local environmental battles, they focused their attention on the threat to Amchitka. Both men were increasingly alarmed at the prospect of further bomb tests on the island. The question was how to publicize the issue in the USA, where people were largely unaware that there had even been a Canadian protest.

In Bohlen's words: "Marie and I were frustrated with the Sierra Club, because they weren't doing anything about the nuclear weapons test, and in the United States [the media] weren't even reporting the action by the students here. Marie said: `The Quakers in 1958 tried to sail a ship near Bikini Atoll in the South Pacific protesting the atmospheric testing of H-bombs. The name of the ship was the *Golden Rule*. They were arrested in Hawaii before they got to the site, and that made all kinds of national news…Why don't we get a ship and take it up there?'

"I liked the idea. Then the phone rang. Some reporter wanted to know what was going on in the environmental movement - looking for something to write about. I said, 'My wife and I were just sitting around here talking, and we thought it would be a helluva good idea to take a ship up to the Aleutian Islands and protest the bombing.' The next day, there it was in the newspaper."

The headline, on February 9, 1970, said, "Sierra Club Plans N-Blast Blockade". The Sierra Club in America objected to this new venture by its Canadian affiliate, preferring a more neutral role as simple conservationists. So Bohlen and

***Border Protest** Hundreds of Canadians gather to call for an end to the scheduled US nuclear test programme.*

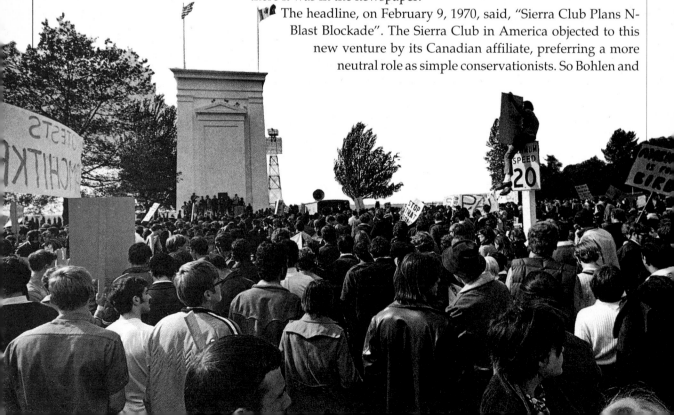

Stowe, together with a young law student from the University of British Columbia, Paul Cote, formed the Don't Make A Wave Committee, borrowing the slogan from the border demonstration in 1969. Its sole purpose was to stop the Amchitka blast.

Accompanied by Patrick Moore, an ecology student at the University of British Columbia, Bohlen soon began attending public hearings in Alaska in an effort to persuade the AEC to stop the detonation of Cannikin. But the key objective was to find a protest vessel with which to "bear witness" at Amchitka. Cote, an expert sailor, began the search for an affordable boat, while Stowe dedicated himself to raising the money to pay for it and finding a crew.

It was at one of the planning meetings, held in the Unitarian church on 49th Avenue and Oak in Vancouver, that the committee found a name for the group that was succinct, and yet so expressive that it generated its own energy and provoked interest everywhere.

According to Bohlen: "The 'Don't Make A Wave Committee' was a lot of words that didn't mean much. People didn't really relate to it, didn't know what it meant. So the group was trying to think of something that was more generic, that people could understand."

Accounts differ as to exactly how the name was put together, but no one disputes that it was Bill Darnell, a young Canadian social worker, who came up with the dynamic combination of words that bound together concern for the planet and opposition to nuclear arms in a forceful new vision that would inspire some of the most effective environmental protests of all time.

"Greenpeace" was born.

Founder Members *In 1970, Jim Bohlen (left), Paul Cote and Irving Stowe formed the Don't Make A Wave Committee, which later became Greenpeace.*

The First Badge *The symbols for ecology and peace are linked to form the first Greenpeace emblem.*

BEARING WITNESS

THE PROTEST BOAT *was to be the* Phyllis Cormack, *an aging 24-metre (80-foot) halibut seiner named after the wife of her owner John Cormack. She would be chartered for six weeks at a cost of $15,000. The* Cormack *had more than her fair share of mechanical troubles. The depth-sounder had to be hammered with a fist to get it working; the anchor winch was worn out; the fuel tanks were damaged by rust; the reversing gear was badly worn and the engine was in appalling condition. But the hull was seaworthy, and having spent 13 months trying to find something better, Bohlen, Stowe and Cote felt they had no choice. This was the boat they would sail against the bomb. She would be dubbed* Greenpeace *for the voyage and would leave on September 15 – little more than a fortnight before the date set for the test at Amchitka. A benefit concert, featuring Joni Mitchell and James Taylor, raised $17,000 towards the cost of the voyage, and the Palo Alto Society of Friends in California sent a whopping $6,000. But most of the money came in dribs and drabs: a dollar sent from a sympathetic follower here, a donation from a fellow environmental organization there.*

Staging Post *A rolling bank of cloud obscures the island of Akutan on the way to Amchitka.*

In the days leading up to departure, the group found itself the focus of attention in the Canadian media. Newspaper reporters, radio journalists and photographers all wanted the latest story on the Amchitka voyage. Even reporters across the border in the United States were at last beginning to take an interest in the protest.

While Irving Stowe dealt with the publicity, Jim Bohlen and a handful of others put all their energies into preparing the boat. More than once they questioned the wisdom of making such a journey in Cormack's battered tub. Described by one crew member as a "floating farmhouse", and barely able to do 9 knots, she would be heading for Amchitka around the autumn equinox. At this time of year, vicious storms begin to form over the Gulf of Alaska and the Bering Sea, producing sudden hurricane-force winds and dangerous riptides that have been known to tear larger, stronger boats in half.

To make matters worse, recalls Bohlen, "Before we went to sail I got a phone call from a person who identified himself as a fisherman. He said 'Whatever you do, don't go out on that ship. It has already been raised twice from the bottom of the Georgia Strait – sunk.' I mean, how are you supposed to feel, the night you are ready to sail?"

IN THE PUBLIC EYE

There were other fears as well, tinged with some paranoia. The committee wondered whether the crew would be contaminated by radiation escaping from the underground blast. Or whether the Americans might bomb them out of the water. Who knew what an irate government might do to a troublesome small boat out of public view?

At least the protesters had the comfort of knowing they were protected by the law. The Quaker boat that had set out for Eniwetok and Bikini years before had been easily stopped. This time the American government would be dealing not with fellow Americans but with Canadians, and any attempt to interfere with them in international waters would be nothing less than an act of piracy.

Furthermore, the boat and her crew would never be entirely out of the public eye. They would have radio communication with the shore and therefore with the media. Journalists and a cameraman among the crew would play a crucial role, recording the events of the journey and sending reports to radio stations and newspapers back home.

A key figure among the journalists was Robert Hunter, a columnist on the Vancouver *Sun*, who had written fervidly about the Amchitka tests. There were also Ben Metcalfe, a veteran of radio, politics and public relations, who was the theatre critic for the Canadian Broadcasting Corporation (CBC), and Bob Cummings, a former private detective now on the staff of the *Georgia Straight*. The photographer would be Bob Keziere, a graduate student in chemistry.

In all, there would be 12 people on board, including Captain Cormack. Nearly 60 years old, Cormack had had 45 years' experience at sea, but this would be the first time he had ventured into Aleutian waters so late in the year. After several years of bad fishing, he was heavily in debt and had little choice but to hire out his boat, even if it meant that he would have to navigate to a nuclear test zone.

Protest Sign *George Korotva (top) helps Robert Hunter (left) and Captain Cormack as they fix the Greenpeace name to the bridge of the Phyllis Cormack.*

Seasoned Hands *John Cormack (left) and engineer Dave Birmingham took no part in the media campaign and weren't much impressed by it. Keeping the old boat going was a full-time job. It kept them together – and apart from the rest.*

Men of the Media *Robert Hunter (left) of the Vancouver* Sun *and Ben Metcalfe of CBC on the bridge.*

66 *As she stopped talking, the old woman and the boy looked to the east and they saw a great rainbow flaming in the sky where a thunderstorm had passed.*

'The rainbow is a sign from Him who is in all things,' said the old, wise one. 'It is a sign of the union of all peoples like one big family. Go to the mountaintop, child of my flesh, and learn to be a Warrior of the Rainbow, for it is only by spreading love and joy to others that hate in this world can be changed to understanding and kindness, and war and destruction shall end!' 99

Willoya and Brown, *Warriors of the Rainbow* (1962)

Led by Jim Bohlen, the rest of the crew would be: Patrick Moore, the ecology student; Bill Darnell, who had come up with the Greenpeace name; Dr Lyle Thurston, a medical practitioner; Terry Simmons, a cultural geographer and one of the founders of the Sierra Club of British Columbia; an engineer named Dave Birmingham; and a political science teacher named Richard Fineberg.

Stowe would coordinate political pressure on shore. He did not go on the journey because of a problem with seasickness. (Three years later he would die from stomach cancer.) Cote, too, would stay behind as he was to represent Canada in an Olympic sailing race.

The boat was now as ready as she would ever be. As reporters from newspapers and television arrived to watch, the last of the supplies were loaded into the hold. There had been a problem getting insurance for the journey – an attempt by the Canadian government, it seemed, to prevent the protest – but that and a myriad other difficulties had been overcome, and Prime Minister Trudeau ended up trying to contact the crew to tell them that he had asked the United States to stop the test.

"THE GREENPEACING OF AMERICA"

At 4 p.m. on September 15, 1971, hoisting a green triangular sail carrying the peace and ecology symbols, the *Phyllis Cormack* – now *Greenpeace* – gave a long, loud toot and rumbled away from the dock. After nearly two years of planning and labour, the very first Greenpeace mission was finally under way – the first in what was to be a string of increasingly famous ventures on the high seas.

On September 16, for a morning commentary on CBC radio, Metcalfe reported from the boat: "We Canadians started the Greenpeacing of America last night, represented by 12 men in an 80-foot halibut boat on our way to the United States nuclear test island of Amchitka in the far Aleutians. Our goal is a very simple, clear, and direct one – to bring about a confrontation between the people of death and the people of life. We do not consider ourselves to be radicals. We are conservatives, who insist upon conserving the environment for our children and future generations of man."

With a tape recorder playing everything from Beethoven to the Moody Blues and with the captain scrutinizing his unconventional crew from the wheelhouse, the *Phyllis Cormack* chugged through the Strait of Georgia, between Vancouver Island and the mainland, bathed in the warmth of an Indian summer. Although the engine broke down briefly on the second day, the old boat was soon well on her way north.

At Alert Bay, a Kwakiutl Indian village on Cormorant Island, the crew went ashore for a special blessing and a gift of coho salmon. They were invited to stop on their return journey, when their names would be carved on the Kwakiutl totem pole. As Robert Hunter recalled, it was no small honour, and seemed to strengthen the "vague affinity" most of the crew felt with the Indians.

Hunter himself had taken on board a copy of *Warriors of the Rainbow*, a small volume of Indian myths and legends that had been given to him a few years before by a wandering dulcimer-maker. "Read this," the stranger had told him. "It'll change your life."

Written by William Willoya and Vinson Brown, the book contained a 200-year-old prophecy that seemed particularly relevant to the men on board the *Cormack*. There would come a time, predicted an old Cree woman named Eyes of Fire, when the earth would be ravaged of its resources, the sea blackened, the streams poisoned, the deer dropping dead in their tracks. Just before it was too late, the Indian would regain his spirit and teach the white man reverence for the earth, banding together with him to become Warriors of the Rainbow.

As the *Cormack* pulled away from Cormorant Island, Hunter passed around his copy of the book. "Predictably," he wrote later, "the older men were less impressed than the youngsters. But rainbows *did* appear several times the following day and it all *did* seem somehow magical as we chugged through a maze of inlets and channels, sounds and bays."

Leaving the shelter of land behind her, the *Cormack* soon found herself in heavier seas, being tossed by the rising swells as if she were a wooden barrel. By the time the boat reached the Gulf of Alaska the sea had turned the colour of granite, and most of the crew had taken to their bunks suffering from bouts of terrible sea sickness.

As they passed through Unimak Pass, the crew sought permission from the US Coast Guard to put in at Dutch Harbor, a naval security zone in the Aleutians, to stock up on supplies and fuel. Then they planned to move to a point just outside the 5-kilometre (3-mile) territorial limit at Amchitka. There, with cameras and tape recorders,

The First Crew From the left: (top) Hunter, Moore, Cummings, Metcalfe, Birmingham; (bottom) Fineberg, Thurston, Bohlen, Simmons, Darnell, Cormack.

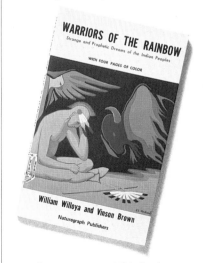

Prophetic Dreams This book was a source of inspiration to Hunter and the younger crew members.

Distant Amchitka *The protesters' goal lies at the tip of the Aleutian arc. Despite valiant efforts, neither of the Greenpeace ships reach the island.*

Weather-Beaten *The sea-chart (background) used on the voyage is torn and coffee-stained after the momentous journey.*

❝ *The Aleutian Islands came up out of the mists – snag-toothed and bleak-looking… We could see froth on white-gummed beaches and hear the distant flush and whoosh of breakers. With cracked chunks of cliff coming into focus like a photo developing… the feeling emerged that we had actually been travelling in a Time Machine back to the beginning of the planet, when volcanoes were still chucking up blood flows of lava. Dankness and cold came from invisible fleets of icebergs in the Bering Sea. The mood could only be described as spooky. We were, after all, entering the most geologically unstable area on earth, epicenter of the Great Alaska Earthquake of 1964. And, of course, an atomic bomb named Cannikin was soon to roar…* ❞

Robert Hunter, *Warriors of the Rainbow* (1979)

the *Cormack*'s crew would "bear witness", in Quaker tradition, to the underground blast – and, through the media, they would make all of Canada bear witness too.

Refused entry to Dutch Harbor because it was a military base, the *Cormack* instead put into waters off the island of Akutan. The crew had received a message over the ship's radio that the test had been delayed, but no one knew for how long. Dismayed at the prospect of having to wait, perhaps for a whole month, Bohlen, Simmons and Metcalfe made repeated attempts to find out exactly when the bomb would be detonated. The most precise date they were given was somewhere between the middle of October and early November.

THE TURNING POINT

After much discussion about what they should do next, they at last decided to leave Akutan on a scouting mission to Amchitka. But, on September 30, the *Cormack* was approached by the Coast Guard cutter USS *Confidence*. The commander came aboard and announced that the *Cormack* was under arrest. The crew had failed to notify customs officials of their arrival in Akutan, and were ordered to the Shumagin Islands – away from Amchitka – to clear customs there.

Behind the commander's back, the crew of the *Confidence* handed the protesters a cablegram. Signed by 17 sailors, it read:

> DUE TO THE SITUATION WE ARE IN, THE CREW OF THE CONFIDENCE FEEL THAT WHAT YOU ARE DOING IS FOR THE GOOD OF ALL MANKIND. IF OUR HANDS WEREN'T TIED BY THESE MILITARY BONDS, WE WOULD BE IN THE SAME POSITION YOU ARE IN IF IT WAS AT ALL POSSIBLE. GOOD LUCK. WE ARE BEHIND YOU ONE HUNDRED PERCENT.

It was a remarkable gesture of support, for such an action almost amounted to mutiny. When the sailors left, their pockets bulged with posters, books and "peace" flags given to them by the Greenpeace crew.

The arrest was a turning point. Besides diverting the *Cormack* from her course, it came at a crucial time in terms of morale. No one could find out just when Cannikin was going to explode, and the tension of waiting was becoming unbearable. Fights were breaking out every day. The crew were all tired, and those who had taken leave of jobs in Vancouver were beginning to wonder when they would ever get back.

The weather was worsening all the time, the sea was hostile, and the waves crashing onto the deck turned into a grey slush. If the test were delayed much longer it would be November before they reached Vancouver and that, said Cormack, would be crazy and dangerous.

On October 12, after an argument that went on late into the night, the crew – with four dissenting votes – elected to retreat.

On the way back they were invited to dock at Kodiak Island, where a banquet sponsored by the city council honoured their bravery. Halfway between Juneau Island and Ketchikan, Alaska, a whale came within 60 metres (200 feet) of the boat, as if to boost their morale. And at Alert Bay they were greeted by 40 Kwakiutl Indians who anointed them and made them brothers of the Kwakiutl people.

Through their disappointment and sense of defeat, the Greenpeace crew had trouble seeing that the battle had, in a way, been won. The voyage had been front-page news in Canada, and Ben Metcalfe's press releases had won Greenpeace support throughout the country. Robert Hunter had also proved expert at aiming media artillery, and the events of the journey – especially the arrest of the *Cormack* and the "mutiny" of the Coast Guard crew – had done what the border blockades had not: they had gained the story at least a mention in the American media.

Such was the goodwill towards Greenpeace that, back in British Columbia, Stowe had been able to raise the money for another, faster ship – a 47-metre (154-foot) converted minesweeper, the *Edgewater Fortune*, capable of doing 20 knots. She headed for Amchitka as the *Cormack* was returning home. The two ships crossed paths near Vancouver Island and four of the *Cormack*'s crew joined the new protest ship for a second try at Amchitka and Cannikin.

From the beginning it was a race against time. Her 28-member crew, provisions and plans were scrambled together in a frantic 24 hours after President Nixon set a deadline of November 4 for the test.

> ❝ *Another expression of public concern was delivered to the White House Thursday – a protest telegram half-a-mile long containing an estimated 177,000 names of Canadians.*
>
> *Western Union, which had been receiving the telegram for four days, said it believes it is the longest message it has handled in its history anywhere in the US. The brief text atop the names said:*
>
> *'As your neighbors, we consider your action in approving this test incomprehensible…You are playing Russian roulette next door to where we live. We ask you in the name of sanity and common sense, to stop it now.'* ❞

Vancouver *Province*,
November 5, 1971

The *Fortune* – rechristened *Greenpeace Too* for the journey – lost time on the 3,200-kilometre (2,000-mile) voyage, when she was beaten back by a storm on her first attempt at crossing the Gulf of Alaska, and when she had to make two stops for fuel and repairs.

Despite her courageous attempt, the *Fortune* was 1,100 kilometres (700 miles) away when, on November 6, 1971, the secretary of the AEC, James R. Schlesinger, gave orders for the bomb to be detonated.

Again there was no earthquake, but there was certainly a tidal wave – of the political sort. There had been so many public protests and demonstrations against the tests, so many threatened strikes and boycotts that it was impossible for the United States to continue testing at Amchitka. After four months of silence, the AEC finally announced an end to the tests in the Aleutians, for "political and other reasons". The voice of Greenpeace had been heard – and the tiny island of Amchitka was once again safe as a wildlife sanctuary.

Sacred Ceremony *In a Kwakiutl long-house, to the scent of cedarwood smoke and the pulsing of drums, the crew are anointed with water and dove feathers and made honorary tribal brothers.*

CHAPTER THREE

JOURNEY INTO THE BOMB

IF THE AMCHITKA voyage was responsible for establishing Greenpeace's name in Canada, another voyage was soon to spread the group's reputation across the world. This new mission would mark the beginning of Greenpeace's longest-running battle, one that would explode 13 years later into a political scandal that nearly brought down a major government.

Atomic Atoll *A mushroom cloud towers over the French nuclear test site at Moruroa.*

At Moruroa Atoll in French Polynesia, France was still testing nuclear weapons in the atmosphere. (Britain, the USA and the USSR had all signed the 1963 Partial Test Ban Treaty and agreed to conduct all future tests underground. France, along with China – the only other countries carrying out nuclear tests – had refused to sign.)

The task of halting these tests would fall to another Canadian, a man who was not a part of the original Greenpeace group and who was not even aware of the events at Amchitka. A former businessman and an expert yachtsman then living in New Zealand, David McTaggart would not only skipper the first voyage to Moruroa, but would also become Greenpeace's shrewdest and most forceful leader.

A championship-class badminton player in the 1950s, McTaggart had always seemed destined for success. At the age of 21, he was running his own construction company and by his mid thirties he was the vice-president of the Bear Valley Development Corporation – a ski resort east of San Francisco. Then, in 1969, disaster struck.

A gas leak caused an explosion at his ski lodge in Bear Valley, and McTaggart had to dig two of his employees from the wreckage, one of them minus a leg. Said McTaggart years later: "Something went out of me then: perhaps the desire to continue fighting for a quality of life I had no taste for; perhaps the rules of the game; maybe the game itself had become meaningless. Whatever the reason, I took what little money I had and boarded a plane for the South Seas. A short while later, *Vega* became my one and only possession."

A 12-metre (38-foot) ketch, the *Vega* had been hand built in the late 1940s in Whangarei by Allan Oram, one of New Zealand's finest boat builders. A strong, sea-worthy yacht, she had large diesel and water tanks and plenty of storage space, making her capable of crossing any of the world's oceans. McTaggart sailed her around the South Pacific, confident she could take any kind of sea conditions. Stopping over in New Zealand, McTaggart met a young woman named Ann-Marie Horne who was to become one of his closest friends.

A RACE AGAINST TIME

It was in 1972 that Ann-Marie's father Gene showed McTaggart the newspaper advertisement that was to change his life.

"My father had noticed this little tiny item about a group called Greenpeace who were looking for someone to sail a boat to Moruroa to protest the French nuclear tests," remembers Ann-Marie. "It was a Vancouver group putting up the plea by way of the Campaign for Nuclear Disarmament here in New Zealand."

The very idea of such a voyage struck McTaggart as fanciful, even bizarre. McTaggart was no activist; he had not given nuclear testing much thought. But having fled the strictures of the business world for the freedom of the sea, he was infuriated that the French intended to cordon off thousands of square miles of international waters around Moruroa – defying the maritime law that gave a nation the right to claim only a 19-kilometre (12-mile) territorial zone.

McTaggart realized that the way to stop the bomb – and challenge the arrogance of the French – would be to sail just outside the territorial

Preaching Peace *A grim-faced David McTaggart stands in the pulpit of his ketch* Vega, *determined to challenge the French at Moruroa.*

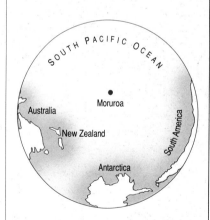

South Pacific *The atoll of Moruroa lies in mid-ocean, thousands of miles from the nearest continent.*

limit. The boat would remain in international waters, so the French could not legally touch her, and they were unlikely to detonate the bomb with a protester right in its destructive path.

McTaggart calculated that the 11,000-kilometre (7,000-mile) trip from New Zealand to Moruroa and back would take three or four months. "To compound the problems, I had heard that the test was scheduled for June 1, which was just over six weeks away. Counting on at least 30 days to get to Moruroa, that meant any fool crazy enough to try it would have to find a crew, stock provisions, and do all the million small things necessary for a voyage of this kind, in two weeks!"

The trip seemed impossible but by now McTaggart was hooked. He had the boat, and he had the time to spare. He telephoned Mabel Hetherington, the 70-year-old honorary secretary of CND in New Zealand, who told him all she could about the French tests. He then contacted Ben Metcalfe in Vancouver, told him that he was planning to make the journey to Moruroa and asked if Greenpeace could pay for a new inflatable life raft and a long-range radio transceiver.

A SOURCE OF FUNDS

Metcalfe was delighted to oblige. He had become chairman of the Greenpeace Foundation after the Amchitka protest when Jim Bohlen, satisfied that no more bombs would be exploding on the island, had decided to pursue other interests. Irving Stowe's careful handling of the funds had left the organization with $9,000 or so after Amchitka, and Metcalfe quickly decided that this money should be tapped to help finance a voyage to Moruroa.

While the Amchitka veteran Dr Patrick Moore prepared for a lobbying trip to New York and then on to Europe, in the hope of getting the United Nations to include discussion on nuclear testing at a

Daring Crew *Nigel Ingram (left) Grant Davidson, Roger Haddleton, Ben Metcalfe and David McTaggart are ready to sail to the test site.*

forthcoming conference in Stockholm, Metcalfe asked to join McTaggart on the *Vega* as a radio operator, and he flew to New Zealand with some $2,500 with which to pay for ship's equipment.

McTaggart was madly dashing about in preparation for this quixotic voyage. As crew he chose: Nigel Ingram, a 26-year-old British navigator who had graduated from Oxford, had been a lieutenant in the Royal Navy, and was an experienced yachtsman; Roger Haddleton, who had served six years as a leading hand with the Royal Navy; and Grant Davidson, a 26-year-old with no sea experience but who was a good cook and badly wanted to make the trip.

The authorities made a determined effort to impede the voyage. Apparently concerned that any such protest might disrupt the nation's dairy trade with France, New Zealand officials made meticulous boat inspections, confiscated a pistol that any serious yachtsman would carry for protection, and threw McTaggart into jail for one night for "smuggling" in wristwatches that he had purchased in Fiji.

But on the last day of April, after making frantic last-minute arrangements and out-manoeuvring the antagonistic authorities, the *Vega* slipped out of Westhaven Harbour, onto a broad reach of dark green sea. The name *Greenpeace III* had been stitched to the sail, along with the ecology and peace symbols.

For the first few days of the voyage they made fantastic progress, powerful winds driving them twice as fast as they had anticipated. But this auspicious beginning was marred by radio difficulties and tension among the crew. The failure of their long-range radio equipment from the outset meant they would be largely unable to communicate with the outside world. The personality problems centred round Metcalfe, who was not only physically unfit and unable to carry out his allotted tasks but was also vying with McTaggart for command of the boat.

TROPICAL FEVER

As a result, McTaggart decided to make a detour of hundreds of miles and put into Rarotonga in the Cook Islands. Here the crew's problems reached a climax. While storms raged, they all went down with a tropical fever and Haddleton was so ill he was unable to continue the voyage. Ingram discovered during a radio telephone call to New Zealand that Metcalfe had filed a press story claiming that the *Vega*'s trip was a decoy mission to distract the French while another boat from Peru sneaked into the test zone. When Metcalfe announced he was leaving for Peru, the rest of the crew felt only relief.

Having put the story straight with the New Zealand press, McTaggart, Ingram and Davidson set out on the last leg of their journey, a 2,400-kilometre (1,500-mile) stretch of lonely and difficult seas. Working round-the-clock watches, the three-man crew fought to reach Moruroa before the June blast.

Against all odds, at 10.45 p.m. on June 1, *Vega* sailed into the forbidden zone and took up her position 32 kilometres (20 miles) from the test site, right in the anticipated path of the fallout. Lingering in the dangerous corridor, the crew held on to their ardent hope that their presence would stop the test and its widespread radioactive debris.

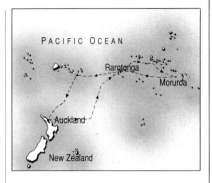

The Mission *The route of the* Vega *on her marathon voyage from New Zealand to Moruroa took her via Rarotonga, where Ben Metcalfe left for Peru and Roger Haddleton was too ill to continue.*

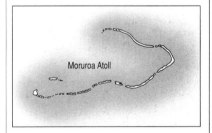

Moruroa *The hollow loop of the atoll, which is made of porous coral resting on a bed of hard, brittle and permeable basalt rock, is the remains of an old volcano. Its name, which means "place of great secret" in Polynesian, was corrupted to "Mururoa" by French naval cartographers.*

Throughout the journey they had broadcast false positions with the aim of putting the French off their scent but, unknown to them, the *Vega* was being constantly monitored by powerful tracking stations in Tahiti and New Caledonia. The next day, a plane passed overhead and a warship sailed close by them. It was like a floating building, mammoth in comparison to the *Vega* and very intimidating. For days the crew of the *Vega* continued their efforts to remain in the fallout zone, despite fierce winds and rolling waves.

The days turned into tension-packed weeks as the three men fought doggedly to maintain the *Vega*'s position. Planes and helicopters flew menacingly overhead. At one point a beautiful grey bird circled above them and headed for Moruroa, oblivious to the firestorm that was soon to erupt. "I wondered what would happen to it when the Bomb went off," wrote McTaggart. "I wanted to shout: Get the hell out of here!"

THE BALLOON GOES UP

It was nearing the middle of June. Positioned just 24 kilometres (15 miles) from Moruroa, the crew of the *Vega* could see the radio towers and bunker buildings of the French base. Following dinner on June 16, McTaggart looked to the horizon and saw what at first he took to be a helicopter but soon realized was something much more sinister.

"It's a balloon!" McTaggart shouted. "A damned, bloody great balloon! They're getting ready to blow off the bomb!"

A scientist in Auckland had warned them that if they could see this balloon, from which the bomb was suspended, they would be too close. At any moment, McTaggart recalled, they expected "walls of heat, unearthly light, shock waves coming across the water like freight trains. Within 15 miles, we would receive third-degree burns – charring of the flesh – and the yacht would be ignited. Even at 20 miles we would be faced with first-degree burns, and the risk of fire was high up to 30 miles. These morbid considerations pursued me …"

Incredibly, the three men decided to force France's hand and move even closer, to a position that would put them under the very rim of any mushroom cloud. They

Harbinger of Doom A gas-filled balloon, seen here before its launch, carried the nuclear trigger device into the sky above Moruroa for detonation.

FORCE DE FRAPPE

The French began their development of a nuclear weapons programme, the *force de frappe*, in the 1950s with the establishment of the Commissariat à l'Energie Atomique (CEA). Between 1960 and 1966, seventeen nuclear tests were carried out in the Algerian sector of the Sahara desert, but following Algerian independence the French moved the tests to the South Pacific.

The first nuclear test in French Polynesia was carried out on July 2, 1966 and was followed in the next eight years by a further 40 tests, all conducted in the atmosphere.

From the very beginning these tests were surrounded by controversy. There were official assurances from the French that "not a single particle of radioactive fallout will ever reach an inhabited island".

Nobel laureate Dr Albert Schweitzer remained unconvinced. In a letter to the leader of the Polynesian Territorial Assembly he wrote: "Those who claim that these tests are harmless are liars."

In September 1966, President De Gaulle visited Moruroa aboard the cruiser *De Grasse* to observe an atmospheric test. Already running late in his busy schedule, he ordered the test to proceed despite the fact that unfavourable winds threatened to blow the fallout towards inhabited islands.

The explosion of this 120-kiloton bomb resulted in radioactive contamination reaching at least as far as Western Samoa, 3,000 kilometres (1,900 miles) away. To this day the French authorities have withheld details of the amounts of radiation to which the Polynesians were ex-

posed by this or any other test. In fact, since 1966 no public health statistics have been published.

As early as 1967, the Territorial Assembly of Tahiti passed a resolution asking the French government to investigate the exact nature of the radioactive fallout, in conjunction with experts from Japan, New Zealand and the United States. No such study took place.

By the following year the first cases of leukaemia had been recorded in the region, as well as cases of changes in skin pigment and a painful rheumatic condition dubbed *la contamine*. A five-fold increase in the radioactivity of rainwater in Fiji was recorded, and radioactive iodine 131 was found in samples of cows' milk in New Zealand, by that country's National Radiation Laboratory.

---

placed wooden plugs beside each of the boat's vents, so they could quickly be hammered into place to seal the cabin. They also decided that if the bomb went off, someone would have to go out on deck to motor the boat from the danger zone. If there was any boat left.

The following day, the *Vega*'s crew transmitted a telegram in the hope that someone would pick it up. It said:

> BALLOON RAISED OVER MORUROA LAST NIGHT. GREENPEACE III SIXTEEN MILES NORTHEAST. SITUATION FRIGHTENING. PLEASE PRAY.

After a day of overcast weather conditions, uncertain whether their message had been received, the crew were galvanized into action by the sight of another French warship. Attempts to get closer to it failed. The following day, it bore down on them and swept by, leaving *Vega* to bob and roll in its eddying wash. Settling just 2.5 kilometres (1.5 miles) from the ketch, the menacing French warship, which they could now identify as the minesweeper *La Bayonnaise*, moved when *Vega* moved, stopped when *Vega* stopped, like a cat playing with its prey.

The "game" continued for two days until, at 6 a.m. on June 21, the French sent an inflatable from *La Bayonnaise* with a written warning to get out of the forbidden zone: the test was imminent.

It was obvious that their presence was causing infuriating delays. The next day, two more ships appeared, one of them a huge cruiser, *De Grasse*. Amid torrents of smoke and a rush of steel, the warships closed in. *De Grasse* roared by and blocked *Vega*'s path, missing her by only 5 metres (16 feet). The other two warships began a game of "chicken" with *Vega*, flanking the tiny yacht and squeezing it between them. "*La*

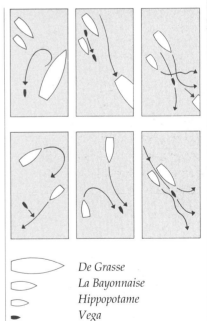

- ⬯⟩ *De Grasse*
- ⬯ *La Bayonnaise*
- ▭ *Hippopotame*
- ➤ *Vega*

Close Manoeuvres *A diagram prepared for the law courts shows the three French warships playing a dangerous game with the* Vega.

Menace at Sea The minesweeper La Paimpolaise *bears down on the* Vega *in a heavy-handed attempt to make her alter course.*

❝ *The two younger men had lived all their lives with the Bomb. Its image had been imprinted on their minds since childhood. They had listened to its roar on the newsreels. It was the single most dreadful image they knew. Previous generations had trembled before visions of hellfire and brimstone, but theirs was the first generation for whom the hellfire was real. It was a hellfire that could be directed and aimed by men. No wall existed that it could not breach. No hole existed that was deep enough for escape. There was no place on earth the fire could not reach. And no father's arm could hold it at bay. Since Hiroshima, no child had been born on the planet who truly knew security.* ❞

David McTaggart with Robert Hunter, *Greenpeace III, Journey Into The Bomb* (1978)

Bayonnaise came in alone and parallel to our port side, a scant 15 yards from us, her huge grey hull rising up and down and the force of her displacement churning the small space of water between us into a maelstrom," wrote McTaggart in his book *Outrage*. "Behind us, the *Hippopotame* was crowding in on our starboard quarter and for a moment or two we rode like that: under full sail, crashing along at about eight knots. Then *La Bayonnaise* began to edge in on us…"

It was a terrifying game, one that nearly caused a collision, and it left McTaggart and his crew badly shaken. The harassment continued for the next eight days. There were helicopters with searchlights at night, a plane buzzing them just over the mast hour after hour, and two ships on their tail most of the time. Through it all, the balloon remained overhead. The bomb had not gone off. But the men on board the *Vega* were too tired to realize that their persistence was working.

The issue was creating dramatic headlines. The London *Daily Express* of June 23 read: "Sail On At Your Peril French Warn Yachtsmen". In Paris, *L'Express* of June 28 led with "Force de Frappe: La Grande Colère Contre La France", and in London, the *Daily Express* confirmed global concern: "Anxious World Floods Paris With Inquiries As The Nuclear Tests Are Said To Have Begun".

Then, at 10.30 p.m. on June 30, McTaggart picked up a French news release on the radio that said Greenpeace had been peacefully escorted out of the area – 11 days before! Now, if something happened, the French could claim they didn't know the whereabouts of the boat.

Still there would be no retreat by *Vega*, which was now so weathered that the mission's name had been washed off the hull. The crew had spent more than 20 days fighting for position around Moruroa after a voyage comparable in distance to crossing the Atlantic. Terrifying though their situation was, they would not stop now.

Deafened and exhausted by two weeks of harassment, the crew thought they heard an explosion at 8.30 the next morning, like a rumble of distant thunder, but they could not be sure. The French were not in fact testing a nuclear weapon but a nuclear trigger device for missiles in submarines. Nevertheless the wind was carrying the fallout to within a mere 24 kilometres (15 miles) of the protest boat.

Vega raised her sails and headed closer to Moruroa, the crew unaware of what they were sailing into. At 10.15 the minesweeper *La Paimpolaise*, ordered to prevent the protest boat proceeding into the fallout, bore down on them. This time, instead of veering off at the last moment, the huge warship collided with the tiny *Vega*, and there was a sickening crunch of battered joints and exploding hardwood.

The *Vega* had been critically injured. Leaking badly, her radio antenna smashed, she was unable even to broadcast a distress call. Left no other choice, McTaggart agreed to being towed into Moruroa. Here the French delivered a *coup de grâce*. During a fine lunch in the company of the Admiral Commandant le Groupement des Expérimentations Nucléaires, the *Vega*'s crew were surreptitiously filmed by three photographers. The pictures, released to the world's press, would almost completely negate the effort of *Vega*'s journey to Moruroa by making it look as if the French and the protest crew were on the best of terms, and the accompanying story claimed that the *Vega*'s misfortune was the result of an ill-judged manoeuvre by McTaggart.

In truth, the *Vega* and her crew had disrupted the French testing schedule. Anxious to be rid of them as soon as possible, the authorities carried out a lightning repair job on the *Vega* to make her temporarily seaworthy, and McTaggart, Ingram and Davidson set out once more on the long haul back to Rarotonga, their morale shattered, their boat leaking badly. They all felt an overpowering sense of failure.

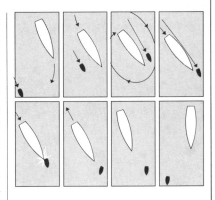

⬭▷	*La Paimpolaise*
▬	*Vega*

Rammed *Another diagram outlines the manoeuvres that ended in a dramatic collision between the minesweeper and the tiny ketch.*

Framed *One of a number of photographs released to the press by the French gives a false picture, implying that the naval officers and* Vega's *crew are on friendly terms.*

CHAPTER FOUR

RETURN TO MORUROA

AS THE BATTERED **Vega** *limped into Rarotonga on July 15, 1972, a defiant McTaggart vowed that the French would not get away with such arrogant behaviour. Leaving Ingram and Davidson to get the boat back to New Zealand, he flew to his old hometown of Vancouver, which he hadn't seen for 15 years. Ignoring those who said it was impossible, he was determined to take a world power to court. "I went back to Vancouver," says McTaggart, "and started my own little war with France."*

"Remember! Moruroa Mushrooms" *A protester dramatizes the dangers of the French tests.*

Disillusioned after three weeks in the city trying to rally support, McTaggart headed for his family's summer home at Buccaneer Bay, on an island off the British Columbian coast. By chance, a yacht carrying Pierre Trudeau, the Canadian prime minister, arrived there and McTaggart had a 40-minute conversation with him, but to no avail. It was clear he could expect no help from that quarter.

In addition, Greenpeace itself had fragmented into a number of rival factions, none of which had the stamina or reserves to fight a long legal case. The news from New Zealand was equally discouraging. Three other protest boats – the *Boy Roel*, the *Tamure* and the *Magic Isle* had tried and failed to reach Moruroa. Worse still, the cost of repairs to the *Vega* had been estimated at $13,000.

McTaggart was not to be discouraged. He had met a lawyer named Jack Cunningham, an expert on marine law, who had agreed to take on his case. News of McTaggart's appearance on national television prompted a message from the Canadian government that they had decided that they would, after all, take action on his behalf.

By this time Ann-Marie had paid her own fare in order to fly to join him and they lived together in spartan conditions in a log cabin on Vancouver Island all through a long and wet winter. McTaggart worked furiously to write a detailed account of the voyage, both because he needed to document it to fight his legal case and because he hoped to raise money by having it published. He wrote letters to supporters and politicians, relentlessly trying to mobilize his case against the French. "My new perception of the world of politics told me one thing clearly: it might be deadlier, but it was the same as both business and sports in one critical respect. If you wanted to win you had to keep pressing, keep pushing, keep sawing away."

When he had almost finished the book, entitled *Outrage*, he found a small local publisher who gave him a $1,500 advance, all but $200 of which he sent to Nigel Ingram in New Zealand to begin repairs on the *Vega*.

It was in April 1973, following the news that the French planned to detonate a hydrogen bomb that summer, that he decided he had to return to the test zone. By now support for his cause was growing, and McTaggart had raised the necessary funds to make *Vega* seaworthy again. An offer of $5,000 from the French on condition that he call off his second voyage was rejected out of hand.

All eyes were now on Moruroa. By this time the opposition to French testing was reaching fever pitch. During 1973 trade unions around the world announced boycotts of French goods, especially in New Zealand and Australia, where French ships, planes, mail and telephone communications were blacked.

"No to Nuclear Testing" A skeleton in Polynesian dress carries the protest message from French illustrator Roland Sabatier.

Warship of Peace *The* Otago *crashes through heavy seas on her way to declare New Zealand's opposition at Moruroa.*

---

66 *The swells were literally the size of small mountain ranges. From the top of one to the top of another was the distance of at least two miles. It took* Vega *fifteen minutes to descend one of those slopes into the valley, ride across the equivalent of three football fields, and begin the long haul up the other side. Coming up over the top, we had such a vantage point that I was quite sure I could see for a hundred miles. For several seconds we would remain poised at the peak, then begin the long shallow descent into the next valley. By the time we were halfway down, the twenty-knot wind would be cut off almost entirely by the vast mass of water behind us so that* Vega *would scarcely move.* 99

David McTaggart with Robert Hunter, *Greenpeace III, Journey Into The Bomb* (1978)

South American nations were upset about radioactive fallout, and in Europe 200 protesters, marching under the Greenpeace banner, set out to walk from London to Paris but they were stopped at the French border and beaten back by riot police. On May 7, police charged several thousand demonstrators near the Eiffel Tower. Physical force seemed to be the government's standard reaction to dissent.

That spring some 25 protest boats were preparing to sail for the test site at Moruroa, leaving from Hawaii, Australia, Fiji, Samoa, Tahiti and Peru. In New Zealand it seemed that anyone who had a vessel capable of making the journey planned to do so. Warships were sent by the governments of both Australia (*Supply*) and New Zealand (*Otago*, later replaced by the *Canterbury*) in what was probably the first ever use of modern military hardware for peaceful protest. Seemingly unperturbed, the French began testing on July 21.

BACK IN ACTION

Soon, the boat that had started it all, the *Vega*, was once more in action. This time she made it back to Moruroa in only 21 days – a phenomenal feat of sailing, through some of the biggest seas that McTaggart had ever seen. He was once again joined by Nigel Ingram, and by their friends Ann-Marie Horne and Mary Lornie.

By August 14, the *Vega* was quite alone. The New Zealand and Australian warships had left the test area, and two other protest boats, the *Spirit of Peace* and the *Fri* had been forced to leave. But on her arrival the *Vega* was immediately spotted by a French aircraft, and the minesweeper *La Dunkerquoise* hove into view.

On August 15, McTaggart saw a small cutter heading in their direction from Moruroa. It met up with *La Dunkerquoise* and was soon joined by their old enemy *Hippopotame*. All three ships closed in on the *Vega* and McTaggart saw that the cutter was towing a dinghy full of men.

"We were in a state of expectation when the French boats started to come around us," remembers Ann-Marie. "It was an impression of boats coming from every direction. This was in the late afternoon. We'd already had a bit of a rehearsal about what we would do if anyone tried to seize the *Vega*. We finally saw this Zodiac dinghy coming across the waves very fast towards us with half a dozen really mean-looking guys. We knew this was really it – they were coming for us.

"We were trying to scamper in any direction we could, but there was no way. Mary and I took our stations with the cameras. Mary was on the side deck and I was on the starboard bow. It all seemed to happen really, really fast – they were coming closer and closer and we started shouting: 'Go away! Go away!' But within a flash they were across the back of the boat and just flailing David."

The next time she saw McTaggart, he was in the dinghy, drenched in blood. The commandos had boarded the yacht, pinioned McTaggart's arms, pummelled him with truncheons, and hauled him off, battering him into unconsciousness. They slammed the truncheons into his head, his kidneys, his spine. One of the savage blows badly damaged his right eye. On board the *Vega*, Ingram too was undergoing a thrashing.

McTaggart's injuries were such that he had to be flown to Tahiti for emergency treatment. The others were taken to the French base at Hao. During the assault, the French commandos had thrown Mary's movie camera overboard and confiscated the camera with which they thought Ann-Marie had photographed the assault. But Ann-Marie had hidden her camera in a cabin as the commandos had clambered on board. Later she was able to smuggle the film of the beatings past the unsuspecting guards by concealing it in her vagina.

Calm Before the Storm Mary Lornie (front), Ann-Marie Horne, David McTaggart and Nigel Ingram on the deck of the *Vega*.

FIGHTING THE FRENCH

Opposition to the French nuclear tests was to prove decisive in shaping the political landscape of the South Pacific region at this time.

It was clear to the Pacific states that Paris was oblivious to the diplomatic protests and that a number of stronger measures were needed to get the message across. These included concerted protests in the UN General Assembly and related bodies; promotion of regional conferences to discuss the issue; imposition of trade embargoes and efforts to establish a nuclear-free zone in the South Pacific.

Such opposition was seen as the main reason for the November 1972 election of Norman Kirk's Labour party in New Zealand, ending 12 years of Conservative rule, and for the election the following month of Gough Whitlam's Labour party in Australia.

Joint Action

The two governments' protest became more substantive when they brought a joint case against France in the International Court of Justice at the Hague.

In the spring of 1973 Australian lawyers approached McTaggart in Vancouver and obtained from him an affidavit concerning the ramming of the *Vega* the previous year.

The two nations' petition included a challenge that France was violating their rights to freedom of navigation, overflight and exploration without prejudice from the nuclear testing.

The case began in June 1973 and a temporary injunction was granted, but France announced that it would not recognize the court's jurisdiction in a matter of "national security".

The case was resumed in 1974, by which time France had already announced its intention to discontinue atmospheric nuclear tests; the court therefore considered that this obviated the need for them to reach a final judgement.

The French naval high command were quick to fabricate a cover story for the press. According to their account, McTaggart, "wanting to throw our sailors back in the sea, [was] himself borne down by their weight and fell into a rubber dinghy alongside the yacht. He injured his eye in the fall when he hit a cleat. Our men boarded his vessel unarmed and without striking a single blow…"

The strongest evidence to counter this was Ann-Marie's photographs. On his release, Ingram had flown with the film to Vancouver, where he had given it to McTaggart's brother Drew together with strict instructions not to release it until later. Greenpeace, however, gently forced him to hand the film over and, two days before McTaggart's return, the group gave the photographs to the press.

The film contained 13 especially damning shots. "The photographs clearly show French commandos pulling alongside the *Vega*," said Greenpeace. "Two of the commandos can clearly be seen to be armed with knives. A rubber truncheon, at least 12 inches long, can be seen in the hands of one of the men as the French swarmed aboard *Vega*.

"The photographs show David McTaggart of Vancouver being hurled into the French dinghy. They also show the French sailors flailing at him in a heap while three other sailors grabbed crewman Ingram, and beat him with at least one clearly visible truncheon."

Outright Lies *These photographs, taken by Ann-Marie Horne, belie the French claim that their men took no weapons aboard the* Vega *and used no violence.*

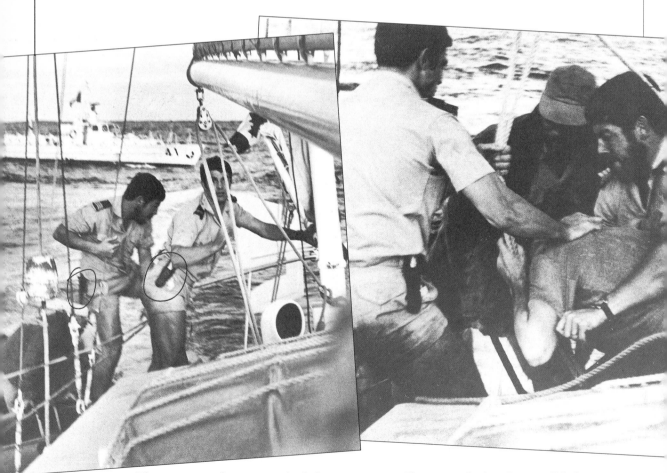

These two men at least are carrying knives.   *The commandos force Ingram off the boat.*

Canadian newspapers now picked up the story. "Film Shows France Told Outright Lie" declared the Vancouver *Sun*. When McTaggart returned from Tahiti he was met at Vancouver airport by an army of reporters and press photographers. By the end of the episode, articles about the savage beatings and the Moruroa voyages had appeared in at least 20 countries and the press clippings filled 22 scrapbooks.

In one sense there was a clear victory. In November 1973, at the UN General Assembly, France announced its intention to stage all future nuclear tests underground after 1974.

McTaggart saw little cause to celebrate, however. For him there were wider issues at stake. The Canadian government had refused to take his case to the International Court of Justice in the Hague which, it later turned out, had been eagerly awaiting it. It would have been the first real test of the validity of the law of the sea since it had been written in the 1600s and might have set a precedent that would affect the exploitation of resources in international waters.

If McTaggart were to pursue the matter further he would have to do it alone, through the French civil courts. On the last day of May 1974, McTaggart, supported by a loan from Ann-Marie and funds raised as a result of a television appeal, flew to Paris to begin what would prove to be yet another long and wearying struggle.

Patched Up David McTaggart, his eye badly damaged by French commandos, recovers in hospital, determined to bring the French to book for their act of "piracy".

McTaggart is beaten and thrown to the deck. *The commandos pick up McTaggart and throw him into their dinghy.*

THE SUCCESS OF THE *FRI*

One of the boats in the "peace flotilla" that attempted the protest journey to Moruroa in 1973, and made it, was the *Fri*.

Her voyage was organized by a group calling itself Peace Media, and the boat's departure from New Zealand on March 23, 1973, was widely hailed in the press. Built in Denmark in 1912, the 32-metre (105-foot) *Fri* was a wooden Baltic trader with a gaff rig and a hand winch, rope and canvas. She had been purchased in 1971 by an American named David Moodie, who upon his arrival in New Zealand was approached by the Campaign for Nuclear Disarmament to join the protest against the testing.

After sailing via Pitcairn Island, *Fri* arrived at the periphery of Moruroa's danger zone on May 25, where she rendezvoused with the *Spirit of Peace*.

On July 15, the *Fri* was resupplied by the charter boat *Arwen*, which arrived with four eminent French activists on board who had come to

Determined *The* Fri *made her presence felt at the test site, despite several setbacks.*

lend their support to the protest: Brice Lalonde, president of Les Amis de la Terre in Paris; Jean-Marie Muller, pacifist; Father Jean Toulat, author of *La Bombe Ou La Vie*?; and, most significantly, General Jacques Paris de Bolladiere, a famous war hero turned pacifist.

On July 17, three crew members from the French warship *Dunkerquoise* delivered an order to keep clear of the security zone. When this was ignored, the *Fri* was boarded, the crew were arrested and their boat was towed to Moruroa. The crew were then flown 480 kilometres (300 miles) north to the French base at Hao, where they began a hunger strike. After eight days they were taken to Papeete for hospital treatment before finally winning the unconditional release of the *Fri* on July 28.

Reunited with their boat, the crew then suffered a further setback when fire broke out in the engine room. Denied landing privileges at Tahiti, the boat had to be repaired offshore before she was able to sail for Moorea, where she stayed for 18 days before heading back to the test zone. She was 580 kilometres (360 miles) due east of Moruroa when the French announced that their 1973 test series was complete. The *Fri* returned to New Zealand to a hero's welcome.

"The court case was being paid for by me personally, and by anyone who would send in $10 or $20," says McTaggart, who eventually had to sell his beloved *Vega* in order to raise more funds.

"I set up a bank account under the name of 'Greenpeace' in West Vancouver. I arrived in Paris with about $200 and I got a little tiny room that cost about $3 a day… I was trying to sue a government that had even tried to block me from coming in. They wouldn't give me a visa. I got on the plane anyway and, basically, I scrounged."

The case took a couple of years to reach its conclusion, during which time McTaggart was working from an office in Paris, receiving invaluable help and advice from Brice Lalonde, president of Les Amis de la Terre (Friends of the Earth), and from a student lawyer, Thierry Garby-Lacrouts, who volunteered his services.

The writ filed by McTaggart claimed $21,000 damages for the ramming and boarding of the *Vega*. On June 17, 1975, he won a substantial victory at the Paris Civil Tribunal when the court ruled that the French Navy was guilty of ramming his boat and that it must pay damages – the amount to be decided by a court-appointed assessor (whose fee must also be paid by the French Navy). On the second and more serious charge of piracy the court ruled itself incompetent to judge, on the grounds that it was outside its jurisdiction, and told McTaggart that if he wanted to pursue the matter he must appeal to another court.

McTaggart persevered but it was not until January 1976 that this aspect of the case was heard, and it was June before a final ruling was given. The court accepted the French government's position that the boarding had been an "exceptional case" and that the decree covering international waters and the conduct of the military were matters of state security and could not be questioned by a French court.

But at the end of the reading of the lengthy verdict, the government procurator made a surprising comment. He said: "It should not be denied that McTaggart may have helped to persuade the French government to decide to choose underground tests in place of atmospheric tests." And, he added: "It is very possible that McTaggart's attitude, reinforced by the reactions of certain countries and certain groups, caused the government of France to think again."

It was a hard-won admission of the success of *Vega*'s voyages. But the issue was far from dead.

TE TUPITA E OE

TE HOE PARAU HAAMARAMARAMARAA POTO NO NIA I TE MAU TUURAA TUPITA ATOMI FARANI

"The Bomb and You" This Polynesian protest poster illustrates the fears of the people in the fallout zone.

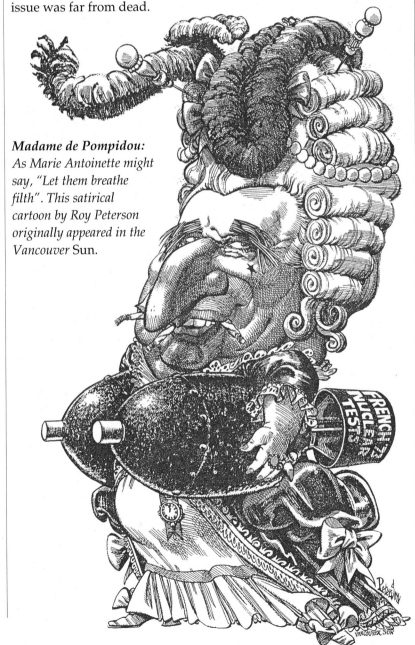

Madame de Pompidou: As Marie Antoinette might say, "Let them breathe filth". This satirical cartoon by Roy Peterson originally appeared in the Vancouver Sun.

66 *Thierry's tiny apartment, where we did most of our work, was only two blocks from the house where Hugo Grotius had lived in the early 1600s when he was commissioned to write a law of the sea to bring an end to piracy.*

That particular work played the most significant part in the case… Describing the sea as something 'which cannot be held nor enclosed, being itself the possessor rather than the possessed', young Hugo Grotius sat in his Paris flat and wrote a law that was to stand unchallenged right up into the twentieth century. Two blocks from where he laboured, Thierry and I now scribbled furiously as we assembled the first test case of the validity of that ancient law. This did something for my sense of perspective. 99

David McTaggart with Robert Hunter, *Greenpeace III, Journey Into The Bomb* (1978)

CHAPTER FIVE

TO SAVE THE WHALE

FROM AMCHITKA TO MORUROA, Greenpeace's main preoccupation up until 1975 was the issue of nuclear testing, but this was to alter radically as the organization gradually became aware of the plight of the great whales, many species of which were being driven to the brink of extinction by the whaling industry. The key figure responsible for the development of this new awareness was Paul Spong, a psychologist from New Zealand who had recently been dismissed from his research job at the Vancouver Aquarium for publicly stating that their captive orca (killer whale) wanted to be free.

Breaching Whale *A powerful humpback splits the surface of the sea in a mighty leap.*

At first the anti-nuclear faction within Greenpeace resisted getting involved in the whaling issue, so Spong and Robert Hunter developed the initial ideas under the banner of Project Ahab.

In 1973 Spong, with Hunter's help, set about raising funds for a trip to Japan, one of the principal whaling nations. He travelled to more than 20 cities there during the first part of 1974, playing whale sounds, showing slides, appearing on television and giving lectures.

Despite Spong's impact in Japan, it was obvious to the Project Ahab people that stronger measures were called for, and they began planning an expedition to confront the whaling fleets in mid-ocean. It was while looking through the photographs of the French pursuing the *Vega* in their high-speed inflatable Zodiacs that Hunter and Spong came up with the idea of using inflatables in the whale protest. Once they had located the whaling fleet, they would lower the dinghies into the water and race off to place themselves between whalers and whales – making it impossible for the harpooner to get a clear shot.

In September 1974, the committee explained their plans to a large gathering of volunteers and asked for their help. Over the next few months people from all walks of life dedicated themselves to the campaign. Wrote Robert Hunter: "It was a fine, if unconventional, blend of human talents and skills. There were… dozens of people who regularly consulted the I Ching, astrology charts, and ancient Aztec tables. Yet for every mystic there was at least one mechanic, and salty old West Coast experts on diesel engines and boat hulls showed up at the meetings to sit next to young vegetarian women. Hippies and psychologists mixed freely with animal lovers, poets, marine surveyors, housewives, dancers, computer programmers, and photographers."

THE NEW GREENPEACE

Following the death in October 1974 of Irving Stowe, the prime advocate of a solely anti-nuclear Greenpeace, and with the resignation of the chairman, a teacher called Neil Hunter, the Greenpeace Foundation virtually ceased to function. In effect, Robert Hunter and the other members of the Project Ahab Committee became the new Greenpeace.

At the beginning of 1975, they rented their first real office, three small rooms belonging to an anti-pollution group on Vancouver's Fourth Avenue, where they set about raising the money to finance the voyage. They organized benefit concerts, sold badges, T-shirts, stickers and posters, auctioned paintings and sculptures, and raffled off a parcel of land given to them by a self-proclaimed Zen shaman.

Finding the boats for the voyage this time round was easy; the *Phyllis Cormack* and the *Vega* were once again called into action. (A retired law professor named Jacques Longini had bought *Vega* from McTaggart and now placed himself and the boat at the disposal of Greenpeace.)

Now Greenpeace had only one more obstacle to overcome. Somehow the group had to locate the whaling fleets, or the two Greenpeace boats would end up combing the entire Pacific looking for them.

The task fell to Paul Spong and, early in 1975, he travelled to Europe, to the Bureau of International Whaling Statistics at Sandefjord in Norway. Presenting himself as a whale researcher, Spong obtained

Whale Motif The Kwakiutl orca emblem decorated the literature of the Project Ahab campaign. Greenpeace was soon to give its support to the new cause.

October 1974
Amchitka pioneer Irving Stowe died of stomach cancer in October 1974, a cruel death considering he had been a vegetarian and a health-food devotee who neither smoked nor drank. Bob Hunter remembered him this way in his column in the Vancouver Sun:

"As I come more and more to believe in the Laws of Karma, Irving's life and energy comes into clearer focus…

"He made a substantial impact on this world, perhaps as much of an impact as could be possibly sought-after outside of the realms of politics, literature and art.

"Although never a professional politician, he was nevertheless a political animal. I saw him as a kind of human bulldozer, whose will, once set in motion, was more like an avalanche than the advance of a single individual.

"No one could say that Irving wasted his time here. He expended himself fully. He contributed precisely as much as he could. When other men were lying back, waiting to see what nightmare would materialize next, Irving was moving like a human whirlwind toward the goal of heading the nightmare off."

charts showing where Soviet and Japanese whaling fleets had hunted in previous years. Assuming the whalers would return to their earlier hunting grounds, Spong calculated that, with the limited range of the boats that it had, Greenpeace's one chance of intercepting either of the fleets would be in June when the Soviets should pass within 100 kilometres (60 miles) of the coast of California.

Fearing that the Soviet fleet might change their schedule, Greenpeace planned to get a two-month head start, which would also allow the campaigners to carry out communication experiments with the whales. So on April 27, 1975, *Phyllis Cormack* and *Vega* set sail from Vancouver. Both boats displayed the United Nations flag, and carried the newly adopted symbol for the whale campaign – the Kwakiutl Indian orca crest. An estimated 23,000 people converged on Jericho Beach, an abandoned airbase near the heart of the city, to cheer them off.

Within days, Japanese newspapers were filled with stories about the protest, and the Japanese government threatened legal action if Greenpeace interfered with its whaling operations. As a means of increasing public awareness of the issue, the campaign was an immediate success.

Robert Hunter coordinated the media coverage, alongside the chief photographer, Rex Weyler, a refugee from the American draft. This time the entire mission would also be filmed, by cameramen Fred Easton and Ron Precious, to make a documentary that Greenpeace hoped to distribute throughout the world.

The *Cormack*'s crew also included: Patrick Moore, Walrus Oakenbough (a nutritionist and an adopted "warrior brother" of the Oglala Sioux), Paul Watson (who had also become an adopted Sioux warrior, named Grey Wolf Clear Water, after taking part in the American Indian

Council of War *Greenpeace's Save The Whale campaigners hold their first meeting* (above) *to plan a voyage against the whalers.*

Deep Affinity *Paul Spong communes with the captive orca at Vancouver Aquarium* (opposite). *Spong's powerful evocation of the whale's intelligence and sensitivity inspired thousands of people to support the campaign.*

WHALE SOUND

The Greenpeace campaigners' decision to spend some time trying to communicate with whales was inspired by Paul Spong's tales of the year he spent studying the captive orca at Vancouver Aquarium. It was there, during a series of experiments to explore the whale's learning abilities, that Spong first discovered the creature's extraordinary sensitivity to sound.

"In the course of fooling around, I started trying sound as a reward and found that the whale would do almost anything to receive sound. I started off producing pure tone signals and then got into bells and crystal goblets.

"I might hold a couple of crystal goblets under the water and gently touch them together with a marvellous ringing sound. The whale would come and sit with the very tip of its forehead almost touching them, or it would back off about a foot and sit there, absolutely still, and slightly twist its head to one side as if to give a different perspective on the sound.

"The first time I played Beethoven's Violin Concerto in D minor to the whale, it proceeded to arch its body so that its head was out of the water on one side and its flukes were out of the water on the other side, and it sprayed fountains of water out of its mouth in time to the music. Its pectoral flippers were beating, slapping, boom, boom, on the surface of the water, or else quivering in the air, just perfectly in time with the music. The flukes were waving gracefully backwards and forwards in the air. It was amazing. It was literally a dance."

Waves of Sound Flautist Mel
Gregory serenades grey whales in
the waters off Vancouver Island.

Action Shots Camera whirring, the
Greenpeace crew (opposite) moves
into position between the factory
ship Dalniy Vostok and one of the
killer boats in her fleet.

Too Young to Die Paul Watson
kneels sadly on the back of a whale
calf slain by a harpoon from the
Soviet hunting ship.

uprising at Wounded Knee, North Dakota, in 1973), Ramon Falkowski
(who had sailed on the *Fri* to Moruroa in 1973), George Korotva (a
Czechoslovakian) and Carlie Trueman of Victoria (both an experienced
scuba diver and an authority on Zodiacs).

For the next four weeks the two boats, operating out of a base at
Winter Harbour on Vancouver Island, made several trips in search of
whales. Following the advice of whale communication expert Dr John
Lilly, the *Cormack* was fitted with underwater speakers and hydro-
phones that would allow Greenpeace to broadcast music to the whales
and to record any response that the whales might make. As well as
electronics experts to operate these sound systems, several musicians,
including saxophonist Paul Winter, flautist Paul Horn and Will Jackson
on Moog synthesizer, went along to serenade the whales.

As it turned out, the whales were not indiscriminate music lovers, as
the first encounter with a pod of grey whales confirmed. According to
Jack Richardson's account in *Playboy* : "The more raucous tone clusters
of the synthesizer and the rock song do not inspire an enthusiastic
response from the greys, which render their judgement by submerging
and reappearing after many minutes at a location far from the source of
the concert. 'They are classicists,' Korotva says, and sure enough,
excerpts from Beethoven's Fifth Symphony elicit a happy response."

On June 1 the two boats went their separate ways. The *Vega* headed
for Long Beach, further down Vancouver Island, to continue serenad-
ing the whales. The *Cormack*'s mission took on a more serious note.

THE WHALE BUTCHERS

Greenpeace had been warned that the Canadian authorities were
feeding the Russians reports of their position. So, in order to put them
off the track, the *Cormack* sailed north for a few days before surreptitiously slipping back to Winter Harbour to refuel. News reached the
crew there that McTaggart had won a partial victory in his lawsuit
against the French Navy, and there was a shipboard celebration the
night before they headed out, on June 18, for their first confrontation.
It was on June 23 that the *Cormack* picked up a radio transmission from
a Soviet "factory" ship, the *Dalniy Vostok*. Its task was to
butcher those whales killed by the harpoon boats – stripping off the skin and feeding great hunks of muscle and
blubber into boilers for rendering into oil.

Within a few days the Soviet ships were in sight, about
80 kilometres (50 miles) due west of Eureka, California.
So was a dead sperm whale. The whalers had attached a
marker beacon to the corpse so that they could come
back for it. The creature was bleeding so profusely that
the water around it had turned a sickening roseate. The
whale also appeared to be undersized. "My God!" cried
Carlie Trueman. "It's just a baby!"

Paul Watson told *Vancouver* magazine: "We have
caught the whale butchers red-handed in the act of
taking an undersized whale. Riding out on a Zodiac, I
leap from the inflatable craft onto the slain whale, its

FOR WHALES OR WHALERS?

Greenpeace's whale campaigns were sparked by the widely perceived failure of the International Whaling Commission (IWC) to protect these magnificent mammals.

Established in 1946 by 14 nations involved in whaling on the high seas, the IWC's stated purpose is "to provide for the proper conservation of whale stocks" and "thus make possible the orderly development of the whaling industry". In reality it was a whalers' club, which watched over the destruction of the whales while its members made enormous profits.

The IWC is structured on three main levels. The Scientific Committee assesses the state of whale stocks (always a controversial matter) and passes its recommendations on to the Technical Committee which, in turn, forwards these and other matters to the annual conference of all IWC delegates. It is here that catch limits are set for the coming season,

decisions being made by a three-quarters majority.

The IWC, however, has no power to enforce its decisions. All members have the right to object to any decision; providing their objection is lodged within 90 days, they are exempt from complying with that decision. Furthermore, an objection extends the 90-day period to 180 days, allowing other countries to object. The US is the only country whose fishery laws enable it to apply effective sanctions. It can close its fishing grounds and block the import of the products of fisheries from any country that "diminishes the effectiveness" of the IWC.

Dramatic Decline

Under the IWC, more whales were killed than ever before. Disregarding the advice of its own scientists, the commission set catch quotas well above those giving a sustainable yield, with the result that whale populations declined dramatically.

The largest whales, which gave the greatest yield of oil, were the first to suffer, and it was only when they became so scarce as to make hunting them uneconomical that the whalers turned their attention to smaller species. As a result, by 1972 the blue whale population worldwide had been reduced from 200,000 to 6,000 and humpbacks had shown a similar decline. Right whales, once numerous, were now very rare, and the populations of Pacific grey, sperm and sei whales had been cut by half.

Faced with this appalling situation, conservationists started to plan in the early 1970s to wrest control of the IWC from the whalers. Their first step was taken at the 1972 United Nations Conference on the Human Environment in Stockholm, where a resolution calling for a 10-year whaling moratorium was passed. It was to take another 10 years of intense lobbying before the IWC accepted a moratorium.

Regrouping The James Bay *and* Phyllis Cormack *lie side by side at their mooring off Sidney in British Columbia*

Tactics Patrick Moore (left) and George Korotva pore over charts of the whaling grounds and plan the Cormack's *course.*

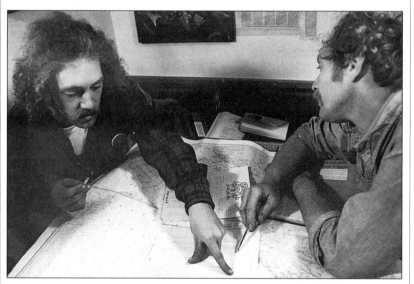

skin warm and oily, the blood flowing from the gaping wound in its side, hot on my hand. I stroke the flipper, reach down toward the vacantly staring open eye and close the eyelid. I am lost and lonely upon the ocean with that dead whale child…"

A harpoon boat bore down on the inflatable, threatening to spray it with a high-pressure hose. The crew immediately retreated to the *Cormack*, fearing that photographic equipment on board the Zodiac would be damaged. But soon the Greenpeace crew headed for an extraordinary confrontation with the giant factory ship.

As two of the Zodiacs buzzed around the *Dalniy Vostock*, with cameramen Easton and Weyler filming the whales being fed into the bowels of the ship, blood gushing from a waste outlet in its hull, the *Cormack* pulled alongside. To the astonishment of the Soviet crew lining the deck, the Greenpeace people took up their guitars and sang anti-whaling songs, then serenaded them with tape-recordings of the songs of humpback whales, played at full volume through the loudspeakers.

Phyllis Cormack then chased after another harpoon boat, the *Vlastny*, which had just unloaded six whales onto the factory ship and was now hunting for more. Soon whale spouts were clearly visible directly ahead of the Soviets. Within minutes the Zodiacs were in the water. In one, Hunter and Korotva raced to position themselves between the harpoon guns and the whales. Patrick Moore fought to keep his Zodiac alongside them while Rex Weyler was frantically snapping pictures. In the third craft, Watson and Fred Easton pounded across the waves to join them.

Suddenly a harpoon was fired just over the heads of Hunter and Korotva, plunging into a whale right next to them, the grenade on the harpoon exploding in the back of the defenceless animal.

The harpoon cable lashed down less than 1.5 metres (5 feet) from Korotva and Hunter. "They didn't give a damn whether they blasted us out of the water or not," fumed Korotva. This close call was captured on film and was soon to be famous. "For the first time in the history of whaling," reported the *New York Times*, "human beings had put their lives on the line for whales."

Although that unfortunate creature died, Greenpeace's actions allowed at least another eight whales to escape the Soviet harpoons.

As the *Dalniy Vostok* moved off toward the horizon, the *Cormack* headed for San Francisco, where the press was waiting along with a huge crowd of supporters and well-wishers.

Every nation that had a newswire, it seemed, picked up Weyler's photographs. TV companies throughout the United States, Canada, Europe, even Japan, showed footage of the Greenpeace crew in front of the *Vlastny*'s harpoon. Journalists demanded interviews, eyewitness accounts. The mission had been an unqualified success.

Within the next couple of years groups calling themselves Greenpeace sprang up from Tennessee to Saskatchewan, their founding members fired

Frontline Film-Maker *A tired but resolute Fred Easton shoulders his camera after another encounter with the Soviet whalers. Film footage of Greenpeace's exploits proved vital in raising public support for the whales.*

Mute Testimony *The tails of two sperm whales lashed to a whaler's hull speak volumes.*

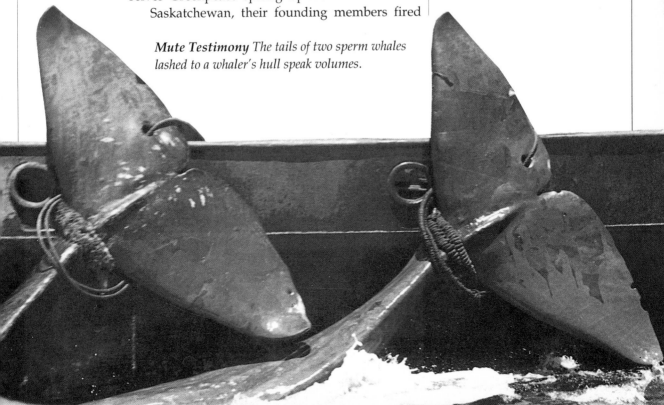

Bloody Wake The butchering of whales aboard the factory ship stains the sea behind it red.

"Whales? Oh No Man, No Whales Aroun' Here" Greenpeace puts one over on the whalers in Randolph Holme's illustration (opposite). *He was a contributor to the* Georgia Straight (below), *an underground paper that chronicled early Greenpeace actions.*

with enthusiasm by Spong and Hunter's lectures. "At least a dozen other independent bands of eco-freaks, each calling itself Greenpeace, spontaneously opened offices in places like Honolulu, Boston, Los Angeles, Seattle and New York," said Hunter, who had assumed leadership. "In Canada we expanded into Lethbridge, Edmonton, Winnipeg, Thunder Bay, Toronto, Montreal, and even, briefly, Newfoundland. Within British Columbia Greenpeace groups sprang up in dozens of little ecology-conscious towns and ports."

Although Hunter reported that there were 10,000 supporters by 1976, many of the "offices" were simply an interested individual who lent a telephone and a room for occasional meetings. In reality, Greenpeace was still a fledgling organization, with only 30 workers at its core. But in San Francisco a solid group was beginning to form, orchestrating fund-raising campaigns that enabled an office to open there in 1976. The USA was waking up to Greenpeace's existence.

In Vancouver, the main Greenpeace office was $40,000 in debt and had to bring some professionalism into its operation. An official board of directors was formed, with Hunter as president, and Moore, Korotva and Rod Marining (a Greenpeace member since the Amchitka days) as vice-presidents in charge of policy, communications and operations respectively. They brought in financial advisers and book-keepers, hired a full-time office manager, organized computerized mailing, set up a merchandizing shop and started their own newspaper.

By early summer 1976 Greenpeace had raised enough money to finance a second mission against the whalers. It was launched from Vancouver on June 13, this time aboard the *James Bay*, a 47-metre (153-foot) retired Royal Canadian Navy minesweeper. She had been chartered under a complex arrangement whereby Greenpeace undertook to put in $50,000-worth of work to make her seaworthy. Ironically, she was similar to the boat that had rammed David McTaggart in 1972, except that the *James Bay* bore a freshly painted rainbow on her bow.

The *James Bay* was fast enough to keep up with the whalers, and had room for 36 crew. "If Russia and Japan decide to whale any longer," warned Robert Hunter, "they will have to do it over our dead bodies."

WARRIORS OF THE RAINBOW

They launched from the site of the United Nations "Habitat" environmental conference, which they had managed to get extended so that its final day would coincide with the start of the voyage. There was another huge concert to send them off. At the dock, Fred Mosquito, a Cree medicine man, told the crew: "You *are* the Warriors of the Rainbow."

After stops in San Francisco and Portland, Oregon, the Greenpeace boat went out to meet the whalers. Again the Soviet fleet was centered around the factory ship *Dalniy Vostok*. Interfering with a hunt in progress in July, about 2,250 kilometres (1,400 miles) southwest of San Francisco, four Greenpeace inflatables formed a barrier between the Soviet boats and a family of small whales, making it impossible for the harpooners to hit their targets. At one point, a Zodiac drove on to the back of a dead whale, and the crew were almost killed as they were hauled out of the water with the pitiful carcass.

Bodyguard *Greenpeace crew member Mike Bailey weaves across the Soviet killer boat's line of fire to protect a fleeing pod of sperm whales from the harpooner.*

66 *We had stirred up a hornets' nest. The Dalniy Vostok itself was steaming at full speed towards the scene. From all over the horizon, harpoon boats had changed course and were likewise throwing gusts of smoke into the sky as they converged toward us. Everything was in kaleidoscopic motion: twelve whales, three Zodiacs, one halibut seiner, ten harpoon boats, and a looming, full-steaming factory ship like a startled rhinoceros charging to the centre of a disturbance.* 99

Robert Hunter, *The Greenpeace Chronicle* (1979)

For a total of 10 days the Greenpeace ship was either in confrontation with the Soviet fleet or in close pursuit of it. At one point they were able to cruise alongside the *Dalniy Vostok* and, through loudspeakers, appeal to them in six languages to stop whaling. It was on the way back northward that the crew of the *James Bay* spotted a submarine monitoring them. The craft tailed them on and off for more than a week, but not once did it come close enough for them to identify it, and they never discovered which nation was watching them so closely.

Of the Japanese whalers there was no sign, despite a report from the US Coast Guard that a Japanese whaling fleet had been spotted near the Hawaiian Islands. With the unofficial help of sympathizers in the US Coast Guard, Greenpeace conducted an aerial survey of some 440,000 square kilometres (170,000 square miles) of ocean from Oahu to Midway, but after six days' searching, the reconnaissance team found no trace of the Japanese. Then, says Robert Hunter, "a report from Honolulu indicated that the US ambassador had suggested to the Japanese ambassador that with anti-whaling feelings running so high, it would be considered unfortunate if there were to be a confrontation between whalers and anti-whalers in nearby waters. Not only the whaling fleet had been asked to withdraw, but all Japanese fishing boats as well."

According to Paul Spong, during its second anti-whaling campaign Greenpeace directly saved 100 whales and, by keeping whalers from normal hunting grounds, indirectly saved another 1,300 or more. Such success was encouraging, but the battle was far from won. The fight to save the whales is still going on more than a decade later.

Circle of Life *Through the ages, the motif of two whales forming the infinite cycle of nature has symbolized the sea-faring Kwakiutl people's desire to live in harmony with the natural world. When they offered their symbol to Greenpeace, the campaigners were honoured to accept the gift.*

BLOOD ON THE ICE

IN THE PERIOD between the two whaling voyages, Greenpeace took up another cause – that of the Newfoundland harp seal pups. Each year hundreds of thousands of harp seals were being slaughtered, mainly by commercial sealing fleets from Norway and Canada. The hunters descended on the pups in February and March, when they were just a few weeks old, clubbing them on the head and stripping them of their fur right there on the ice.

"Over My Dead Body" A special Greenpeace
issue of the comic Slow Death.

B y the mid 1970s, intensive hunting had drastically reduced the seal population. Native peoples and early settlers had taken adult seals for their meat, skin and oil, but the commercial sealers, who operated for many years without quotas, took the pelts of the young seals to make coats, gloves, fur covered trinkets, ski boots and other "luxury" goods.

Public attention had first been drawn to the seal slaughter years before when, ironically, the seal hunt was featured in a television film promoting tourism in Quebec. A sequence meant to portray the ancient struggle between man and nature, showing newborn seals being clubbed to death, shocked viewers across Canada and the USA.

Among those viewers had been Walrus Oakenbough and Paul Watson, who had discussed the hunt on board the *Phyllis Cormack* as she made her way from San Francisco to Vancouver on the final leg of the first whaling trip in 1975. Their idea was to travel to Newfoundland and attempt to stop the massacre by spraying the pups with a harmless green dye so that their pelts would be of no use to the hunters. If their pelts were worthless, the pups' lives would surely be spared.

On March 2, 1976, the first Greenpeace expedition to save the seals left Vancouver to travel cross-country by train to Nova Scotia, by ferry to Newfoundland, and by road up the length of the island to the port of St Anthony. There they were to be met by two helicopters, chartered to take them to the ice floes where the hunt took place.

It was late winter and snowstorms were raging. At several points the van swerved off the narrow and slippery road to St Anthony, where the temperature had plunged to -20 °C (-4 °F). As if the treacherous weather were not enough to contend with, Greenpeace was met by a gang of angry Newfoundlanders, blocking the road into town. As the Greenpeace van drew slowly to a halt, the crowd pressed against it, trying to push it over. But the violence was short-lived, and a meeting was arranged for later that evening, enabling each side to put its case.

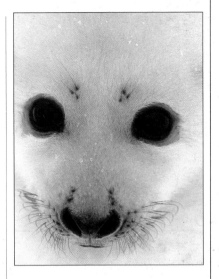

Silent Appeal The pup of the harp seal, Pagophilus groenlandicus, is born in early spring and has a downy white coat for only two weeks. The pups are clubbed to death before they are even weaned off their mothers' milk.

Bloody Harvest A heap of harp seal pelts lies on the ice awaiting collection. To kill the seals, hunters strike them on the head with a hakapik, a primitive spiked cudgel.

A Stain on the Ice Blood from a parcel of pelts marks the route to the sealing ship.

Badge of Protest *Greenpeace's campaign drew public attention to the inhumane harp seal kill.*

The Newfoundlanders were not the only ones annoyed at Greenpeace's interference. The Canadian government had hastily drawn up a new law making it illegal to spray seals and had banned anyone from moving a pup or placing themselves between a seal and a hunter. The activists now faced the prospect of ending up in jail. "It seemed wise," recounts Hunter, "to change tactics – rapidly."

At a raucous meeting of 400 people, Hunter announced that, as a gesture to the residents of St Anthony, Greenpeace had decided to drop the plan to spray the seals and would hand over the dye the following day. The decision was controversial and caused uproar among Greenpeace supporters who saw the concession as an act of surrender. The Greenpeace office in Vancouver was inundated with calls from people demanding their donations back. But Greenpeace had no intention of surrendering. Its new tactic was to focus attention on the Norwegian commercial sealing fleets.

By now St Anthony was overrun with reporters and film crews from Germany, Canada and the USA. When David McTaggart arrived from France he had a further television crew and a photographer from an international news agency in tow.

On March 15, the Greenpeace helicopters set out from the base camp on Belle Isle, some 50 kilometres (30 miles) north of St Anthony, ready for the first encounter with the Norwegians. Prohibited from landing within 800 metres (half a mile) of the seals, the protesters had to make their way on foot across the shifting ice to the hunting grounds.

A HUMAN SHIELD

As they approached, the air was filled with desperate wails and screams, the mother seals standing by helplessly as their offspring were brutally clubbed and skinned. (The blood had been visible even from a helicopter more than 600 metres [2,000 feet] above the ice.) Greenpeace member Al "Jet" Johnson wrapped himself around a pup, shielding it bodily from a sealer with his raised hakapik. Time and again that day the same action was repeated. Once more, Greenpeace had succeeded in placing itself between the hunter and the hunted.

As night fell, the helicopters whisked the protesters back to Belle Isle. There the storm worsened, the wind gusting to nearly 160 kilometres per hour (100 mph), forcing Greenpeace to evacuate the base camp and return to St Anthony to wait for three days until the weather improved, before resuming the protest in earnest.

Back to Base *Arriving by helicopter, three Greenpeace members return from their protest on the killing fields.*

Blown Out *Strong winds and driving snow on Belle Isle (above) force the Greenpeace team to abandon base camp – and the protest – for three days.*

Act of Defiance *Disregarding hastily drawn-up and draconian Canadian laws, Paul Watson (left) takes a seal pup in his arms to protect it from the sealers.*

The morning of Friday, March 19 brought light winds and good visibility, but when the Greenpeace team returned to Belle Isle they found that the storm had swept their tents, personal belongings and equipment off the island and out to sea.

Using the helicopters, they flew 130 kilometres (80 miles) south of their last encounter with the sealers and landed on the ice. The sealers were at work and bloody pelts lay everywhere. The only course of action for Greenpeace seemed to be to try to prevent the sealing vessel from moving further into the ice. Paul Watson takes up the story:

"Bob Hunter and myself find a seal twenty feet before the monstrous scarlet ice-crushing bow of the *Arctic Endeavour*. We hold our position with our backs to the ship.

"On the starboard side of the vessel, a crewman is busily attaching and securing winch lines to a bundle of blood soaked pelts. He yells out to us, 'Ya betta move b'yes, the ole man ain't one ta tink twice bout running ya inta the ice.' Bob yells back, 'Tell the old bastard to do what he wants, we're not moving.'

"The ship backs away. We think we've won this round. Then the unexpected. The *Arctic Endeavour* plunges forward picking up speed. Still we look ahead. We feel her coming. We hear her coming. The vibrations of the powerful diesel engines disturb the chilly air and tingle the soles of our feet through our boots. The ice trembles and cracks. Blocks of chunky ice tumble forward before the bow and nudge our feet.

"The crewman on the ice screams to the ship's bridge. We can make out his words clearly. 'Stop 'er Cap, stop 'er, the stupid asses ain't a moving.' The engines are cut and reversed. The ship slowly grinds to a halt, five feet behind our backs.

"I pick up the baby whitecoat to remove it to safety. My way is blocked by a uniformed fisheries officer. He takes my picture. He pulls some papers from his pocket and begins to read me the amendments to the Seal Protection Act. 'Section 21 (B) states that it is a Federal offense to remove a seal from one location to another. It is an offense to pick a live seal up from the ice. You are in violation of this regulation.' He leaves me no choice. I ignore the law and carry the seal to safety."

Further trouble with the Canadian authorities was to come. The following day officers from the Fisheries Department claimed that the helicopter pilots were in violation of another of the Seal Protection Act regulations because they had flown lower than 600 metres (2,000 feet) over the ice. On March 21, the Greenpeace helicopters were grounded and placed under the guard of the Royal Canadian Mounted Police. Greenpeace had to pay a bond of $10,000 on each helicopter before the aircraft were freed pending a trial.

Such petty harassment by the authorities was to characterize the campaign in years to come, but Greenpeace was not to be deterred. Although the group's actions saved only a handful of seals that year, it had succeeded once again in drawing attention to its cause.

SPREADING THE WORD

The "active membership" of the group was now put at 8,000, with 13 "very active" branches around the world, as well as 28 "heard from occasionally", according to the Vancouver *Sun*. Most of these branches were merely temporary outposts that would soon fade away, but the seeds of Greenpeace were rapidly spreading.

By the following year, 1977, the seal campaign had become an international one, including people from Norway, Britain, the United States, and Canada. In the eyes of the media, the high point was the appearance on the ice of the French actress Brigitte Bardot. At least 45 journalists had accompanied Bardot from Europe, representing Denmark, Norway, the UK, Germany and Switzerland.

While Bardot was holding the attention of the media, the Greenpeace campaigners were once again trying to stop the Norwegian sealers. After charging across the ice, its unstable surface shifting beneath his

Save the Seals *Mike's artwork became the UK campaign sticker.*

Lives on the Line *In an heroic action* (opposite)*, Paul Watson (left) and Robert Hunter block the path of the* Arctic Endeavour.

Star Attraction *Brigitte Bardot, actress turned animal campaigner, expresses her concern for the seals' plight by joining Greenpeace protesters at Belle Isle. Her short trip to the ice assured the issue prime time media attention.*

feet, Paul Watson handcuffed himself to a cable being used to hoist a load of pelts on board a seal ship. He was dragged across the ice and repeatedly dunked in the water and was nearly drowned in the process.

"It was like being in the Roman Coliseum," Watson said. "They were yelling 'drown the bastard' when they pulled me on board and laid me face down in the seal pelts so that I got all bloody."

While Watson was on the way to hospital having suffered a dislocated shoulder in the confrontation, the other Greenpeace activists continued their protest, saving individual pups and forcing a sealing ship to retreat, leaving 100 pelts behind on the ice. That year for the first time there was a significant reduction in the number of seals slaughtered. Rivalling the whale campaign in terms of publicity, the seal protest was to become an annual event.

But there was precious little else that was regular. The organization continued to be riven by internal disputes that marred its effectiveness. Hunter resigned as president and was replaced, on April 20, 1977, by Patrick Moore. Watson, who was accused of creating dissension over campaign tactics on the whaling issue, was removed from the board of directors after a long and bitter dispute. (He was later to form his own

Dead Weight *Handcuffed to a bloody bale of seal pelts, Paul Watson is hauled into the air by the sealing vessel's crane.*

conservation group, the Sea Shepherd Foundation.) Captain Cormack, also a board member, found himself in serious debt and faced losing his boat. The Vancouver office continued to have substantial debts of its own. Various splinter groups were creating their own bureaucracies and hierarchies. A man named Gary Zimmerman was elected president of the self-proclaimed Greenpeace Foundation of America, Inc. There were so many power struggles – between individuals as well as branches – that the situation was becoming a nightmare.

Amid this continuing turmoil, however, one strong thread was in the process of being spun. During 1976 and 1977 David McTaggart, still battling against the French government over its piracy and damage to the *Vega*, had been bringing together various activists who would soon form the core of Greenpeace Europe.

Skinned *The carcass of a seal pup lies discarded on the ice* (above).

Role Reversal *This highly personal view of the seal issue* (left) *comes from the pen of the French illustrator Moebius.*

"Protect the Seals" This campaign sticker was produced by Greenpeace in France. The seals issue was to prove a controversial one in Europe.

Predator and Prey *A lone seal pup, not yet weaned and unable to swim, lies helpless on the ice as a sealing ship advances.*

Working from his base in Paris, McTaggart also started to use the offices of Friends of the Earth in London as a mailing address, and soon began talking to two FOE members, Susi Newborn and Denise Bell, about setting up a London Greenpeace office. Coincidentally, Newborn had advertised for a flat-mate and one of the replies was from a Canadian member of Greenpeace named Allan Thornton. He was attempting to bring the annual seal slaughter to the attention of people in Europe, where the skins were being sold.

As Peter Wilkinson, yet another member of Friends of the Earth who joined Greenpeace, recounted later: "In 1977, four people gave birth to Greenpeace in the UK from a borrowed office in London's Whitehall with £800 and a lot of determination."

Meanwhile, in Paris, at a demonstration of trade unionists from France's nuclear power plants, McTaggart had met a young activist named Remi Parmentier – again, a member of Friends of the Earth. Together they made plans to establish a Greenpeace office in Paris, which began operating officially in 1977.

The European volunteers also began discussing ways of joining in the campaign to save whales. Greenpeace knew that an Icelandic whaling fleet would be hunting in the North Atlantic in May of the following year and it was determined to do its best to stop them. In Denise Bell's mind, there was only one course of action: to get a protest boat.

THE WARRIOR SETS SAIL

DURING THE EARLY PART of 1977 Denise Bell dedicated herself to finding a boat that could take Greenpeace into the whale hunting grounds of the North Atlantic. Through an advertisement in a seamen's journal, she found a rusty 418-tonne trawler languishing on the Isle of Dogs in London. Launched in 1955, the 44-metre (145-foot) **Sir William Hardy** *was the first diesel-electric-powered ship to be built in the UK and had been used by the Ministry of Agriculture, Fisheries and Food as a research vessel. Whatever the boat's outward appearance, an engineer confirmed that she would, eventually, make a good campaign vessel. Greenpeace had, however, yet to find the £44,000 to pay for her. It took eight months to raise enough money to make a 10 per cent down-payment; the balance then had to be paid within 60 days.*

Maiden Voyage *The* Rainbow Warrior *sails out under London's Tower Bridge and into action.*

April 1978
A month-long sit-in that blocked the railway line to the Rocky Flats nuclear weapons facility in Colorado, USA (above), led to the arrest of more than 50 people . This was one of a number of demonstrations against the Rocky Flats plant, which manufactures fission triggers for thermonuclear warheads and handles more plutonium than any other facility in the western world.

Graphic Detail *The Kwakiutl orca crest forms the background to an artwork for the* Rainbow Warrior's *campaign.*

Just when hope was running out, David McTaggart arrived with the news that the Dutch branch of the World Wildlife Fund had agreed to make a grant of £40,000 to finance a campaign against Icelandic whaling. (Greenpeace Netherlands was to be established in March that year, principally as a fund-raising operation.)

With the money in hand, the purchase was quickly completed and, in February 1978, the ship was brought across to London's West India Docks. Now a race against time began, as they had just three months to get the boat ready for action. Teams of volunteers arrived from all quarters and began chipping at the rusty hull, overhauling the engines, and fixing the complex and waterlogged electrical system. They cut off some 25 tonnes of winches and fishing gear, renewed the navigation and radio equipment, and turned the laboratory into an extra cabin, bringing the total number of berths to 23. The fish-hold was later converted into a theatre for lectures and meetings.

In that short space of time, the *Hardy* was transformed. Painted dark green, with the word "Greenpeace" emblazoned across her side and the image of a dove carrying an olive branch on her bow, she was rechristened the *Rainbow Warrior*, a name inspired by the same collection of Indian legends that had influenced the first Greenpeace crew.

On April 29, flying both the Greenpeace and United Nations flags to reflect the international nature of her crew of 24, the *Warrior* left harbour and headed up the east coast of Britain. As a prelude to the Icelandic campaign, the boat put into Torness on the East Lothian coast of Scotland to support a rally of more than 3,000 people against the construction of an advanced gas-cooled nuclear reactor.

THE BATTLE BEGINS

After refuelling in the Shetland Islands, she headed out for her first confrontation with the Icelandic whalers. In the battle to save the whales, Iceland was a special target. McTaggart saw the situation like a chess game: "The king is Japan, the queen is Russia and Iceland is a combination of a couple of bishops and a couple of rooks, supporting Japan and Russia in just about all they do. If you stop Iceland, you throw a hook into the whole International Whaling Commission."

With this campaign, Greenpeace hoped to force the IWC to introduce stronger conservation measures and to institute the moratorium on whaling called for at the UN Environment Conference in 1972.

Iceland was a special target for another good reason: it was still killing fin whales, despite the fact that stocks were already depleted. Using four steel-hulled catcher boats equipped with 70-kilogram (150-pound) explosive harpoons, the Icelandic whalers, together with a Norwegian fleet, were expected to slaughter 2,500 whales before the season ended in October.

"We knew it was not going to be a vacation," says Remi Parmentier. "In Iceland the weather and visibility are very bad, even in the summer." There would also be the risk of violence: the Icelanders had been involved in a "cod war" with Britain only two years before, during which they had filled the bows of their trawlers with cement and used them to ram British boats.

Sir William Hardy *The battered trawler lies in dock before her refit (above). Enthusiastic volunteers prepared her for action in just three months.*

Colourful Emblem *A scene of whales, framed in the orca crest, decorates the stern of the* Rainbow Warrior. *The ship is to play a key role in Greenpeace's struggle to end the destruction of these threatened mammals.*

June 25, 1978
A Greenpeace yacht, carrying a banner reading: "Stop Seabrook. Save the Humans", joined a "boat picket" off New Hampshire, USA (above). This was part of the biggest anti-nuclear rally yet held in the United States, in protest at plans to build a twin 1,150 megawatt nuclear power station at Seabrook; the picket formed around the offshore platform used to drill the power station's cooling tunnels. One of the picket boats, the Clearwater, *was skippered by a young man named Peter Willcox. He was later to become the skipper of the* Rainbow Warrior.

As the *Rainbow Warrior* steamed north, two Greenpeace representatives from London and Paris flew to Iceland and staged a press conference in Reykjavik. The sympathetic response they received encouraged them to hold a public meeting at which Mr Loftsson, who was both the owner of Iceland's only whaling station and the Icelandic delegate to the IWC, was present.

The *Warrior* had by now reached the mouth of Hvalurfjord (Whale Fjord), just north of Reykjavik on the west coast of Iceland, from which the four catcher ships were due to emerge for the hunt. Peter Wilkinson takes up the story:

"Despite the exhaustive preparations, fog, heavy weather and an underpowered boat hampered the Greenpeace endeavours for ten frustrating days. The *Rainbow Warrior* could not match the speed of the catchers and dispirited Greenpeace volunteers could only watch as the catchers returned to the land station with their maximum catch of three whales fastened to the bows of the ships before sailing off again, leaving the *Rainbow Warrior* in their wake.

"But Greenpeace were learning. Soon they had managed to decipher the radio signals passed between boat and shore and could predict the area to which the catchers were steaming. On the tenth day the *Rainbow Warrior* was hove-to in an area thought to be the likely hunting grounds and, in the twilight of the perpetual dawn experienced in the Arctic circle in the summer months, the first whales were sighted as they ploughed their way north to the breeding grounds. Within hours, the pursuing catcher ship was spotted. It seemed that this day would see the enactment of what six months ago had been a mere pipedream."

THE FIRST CONFRONTATION

The boat in question was the *Hvalur 9*, which was cruising at reduced speed while the lookout scanned the sea for whales. Suddenly, a few hundred yards away, a pod of whales was spotted and the catcher veered sharply to port. Greenpeace went into action, launching four small inflatables which sped over to intercede between whales and whaler. Allan Thornton describes the scene:

At the Ready *Greenpeace members from the* Rainbow Warrior *test out the inflatables before setting off to intercept an Icelandic catcher ship.*

"We came across the Number 9 catcher at about 8 o'clock in the morning. At that point the seas were moderately heavy, an eight- or ten-foot swell, and he was going right into the wind, making it really bad putting [our] boats out. We were flying right out of the water and coming down very hard. It was like running into a brick wall.

"That day we finally got in front of the harpoons. It was towards the end of the chase aimed at wearing the whales down and we could see the two fin whales breaking right in front of us about 30 yards away. The captain came running down the little walkway they have from the bridge to the harpoon… He was tensing and we really thought he was going to fire. I was really scared. It was a really tense situation."

The confrontation lasted for at least six hours, during which the inflatables, suffering engine troubles and constantly buffeted by swelling seas, found it increasingly difficult to maintain their positions. But, finally, the *Hvalur*, unable to get a clear shot at any of the whales that were now surfacing every few minutes, headed back for Reykjavik. The *Warrior* had had her first taste of victory.

But before the crew had time to celebrate, a second confrontation with another catcher ship took place, this one even longer than the first. Wherever the whaler went in the heavy, rolling seas, Greenpeace trailed it. This pattern was to continue over the course of the next month, as the protesters followed the whalers further and further north. By the time they were forced to head homewards, they had not only succeeded in saving the lives of many whales, but had also drawn whaling to the attention of the Icelandic people and made it a major issue.

A NEW PRIORITY

It was now late June, and the *Rainbow Warrior* was due to head for Spain to stage further whaling protests but, while refuelling in Dublin, she was called into action on an entirely different campaign. Her mission: to intercept a British ship, the *Gem*, which was en route to dump 2,000 tonnes of radioactive waste from the UK in international waters.

It was Greenpeace campaigner Peter Wilkinson who had first noticed a spot labelled "Dumping Area" while studying charts of the North Atlantic. Greenpeace had investigated and had discovered that several European nations had been dumping nuclear waste 1,000 kilometres (600 miles) south-west of the Cornish coast, into a 3-kilometre- (2-mile-) deep trench in the seabed, for the past 20 years. Such ocean dumping had been abandoned by the USA in 1972 on environmental grounds.

Greenpeace had been monitoring the situation for some time but was spurred into action when, on the day before the *Gem* left Sharpness Docks near Bristol, a Royal Navy articulated lorry delivered two yellow drums, which were loaded on board. Information was leaked to Greenpeace that the drums contained spent nuclear fuel rods from submarines. If true, this consignment would breach the London Dumping Convention, an international agreement that prohibits the disposal of highly radioactive materials at sea.

Suspect Load One of the yellow drums believed to contain nuclear submarine fuel rods arrives at Sharpness aboard a naval lorry.

ACTIONS FOR WHALES

During the 1977 summer whaling season, Greenpeace was again in action in defence of the whales, this time with two boats against two Soviet whaling fleets in the North Pacific, and with Zodiacs against land-based whalers in Australia.

On July 17, the *James Bay* sailed from San Francisco and stalked one of the Soviet fleets, consisting of the 180-metre (600-foot) *Vladivostok* and her eight catcher boats, some 1,100 kilometres (700 miles) southwest of Los Angeles, off the coast of Baja California.

When the whalers encountered a pod of eight sperm whales, the Greenpeace Zodiacs spent four hours attempting to frustrate the hunt. Rex Weyler recalls the scene: "One by one, every ship in the fleet joined the hunt until a total of 10 harpoon boats pursued the one pod of whales. Finally the lead boat fired past us, missing us and the whales as well. As the panicked whales were forced to the surface, the boats closed in. Harpoon explosions sounded all around us as we realized our helplessness. They slaughtered every whale in the pod: bulls, females, and calves." If nothing else, the Greenpeace

***Hot Pursuit** The* Dalniy Vostok *is chased by a twin-engined inflatable bearing a banner saying "Nyet" to Soviet whaling.*

cameramen were at least able to capture the incident on film.

A month later, several members of the *James Bay*'s crew (Patrick Moore, Rex Weyler, Bob Taunt, and Russian-speaking Rusty Frank) boarded a Soviet catcher boat tied up alongside the *Vladivostok* and were able to put their case to the Soviet crew.

While the *James Bay* was at sea, the *Ohana Kai* (meaning "Family of the Seas"), a 50-metre (165-foot) ex US Navy sub-chaser, the first ship that Greenpeace had ever owned, was in hot pursuit of the second Soviet whaling fleet, the *Dalniy Vostok* and its attendant catchers, which it tracked down some 1,900 kilometres (1,200 miles) north of Hawaii. The *Dalniy Vostok* was too slow to outrun the Greenpeace ship, which shadowed the fleet for a week, during which time no whales were caught.

One team of Greenpeace crew members, including Paul Spong, Kazumi Tanaka, Nancy Jacks and Dexter Cate, drove their Zodiacs right up the stern ramp of the *Dalniy*

Vostok and climbed onto the deck. They presented the shocked but curious crew of the whaler with gifts of whale badges and anti-whaling literature in Russian.

The *Ohana Kai* carried a nine-member ABC-TV crew equipped with their own Bell Jetranger helicopter, but the best footage came from the Greenpeace cameras aboard the *James Bay*. Film of the voyage and the slaughter of the sperm whales was broadcast nationwide as a one-hour documentary. President Carter made a special request to view it.

In July 1977, the annual IWC conference was held in Canberra, Australia, and was the scene of concerted demonstrations by various anti-whaling groups, including the Whale and Dolphin Coalition and Project Jonah, which had gained considerable support through a campaign of education and political

lobbying. During the conference a 12-metre (40-foot) white inflatable whale floated in the lake next to the conference hall, and finally appeared in the centre itself, blocking the corridor to the Japanese assembly room. In full view of the TV cameras, police and hotel staff had to stab it with knives before it could be removed.

The last English-speaking nation engaged in commercial whaling, Australia had only one remaining whaling station – the Cheyne Beach Whaling Co. in Albany, Western Australia – and the Whale and Dolphin Coalition sought help from Greenpeace to have the station, which was taking 600 sperm whales annually, closed.

Determined to Succeed

An anti-whaling convoy set out from Sydney on August 20 that year to drive 4,800 kilometres (3,000 miles) across the south of Australia to Albany, where they planned to rendezvous with a chartered vessel, the 21-metre (70-foot) *Fabrina*. However, halfway across the Nullabor Plain they broke down and were delayed for three days. During this time the Australian government announced its opposition to the protest and, as a result, the owner of the chartered *Fabrina* withdrew the boat.

Undeterred, the anti-whalers continued their protest using two Zodiacs and in the next month they placed themselves directly in the line of fire. On two occasions harpoons were fired straight over them; in one case the propeller of a Zodiac became snagged on the cable beween the harpoon gun and the dying whale, raising the dinghy out of the water. Video film of these and similar incidents was later shown on the major television networks in Australia.

The protest drew considerable attention, and Greenpeace was able to follow this up with another coup. When the Australian whaling company sent 1,200 tonnes of sperm oil (valued at $1.2 million) by tanker to Portland, Oregon, Greenpeace alerted US Customs and the shipment was refused entry. The whale oil (described as fish oil) was then unloaded into tanks in Vancouver, but Greenpeace informed the local trade unions of the true nature of the consignment and it was held for two months at an estimated cost to Cheyne Beach of $250,000.

These actions helped to turn the tide of opinion in Australia against whaling to the point where 70 per cent of the population wanted to see it banned. In September 1977, Prime Minister Malcolm Fraser announced the setting up of a Royal Commission to inquire into Australia's whaling policy.

In 1978, while the inquiry was hearing evidence, Greenpeace representatives met with the Whale and Dolphin Coalition and an Australian branch of Greenpeace was formed.

The inquiry presented its findings in early 1979 and the government accepted its recommendations in full. As a result, the last whaling station in Australia was closed down, and whaling was banned within its 320-kilometre (200-mile) fishing zone.

"Save The Whales Voyage 1977"
The cover illustration from a Greenpeace campaign leaflet features a cartoon of the James Bay *with playing whales.*

Furthermore, at the IWC, Australia announced its commitment, on both scientific and ethical grounds, to a total whaling ban. Thus, in just a few years, Australia had not only ceased to be a whaling nation but had become the first country to state that whaling was morally wrong.

Putting the Case Paul Spong and other Greenpeace protesters attempt to establish a dialogue with the crew on board the Soviet whaling ship.

Hazardous Cargo Barrels of radioactive waste lie in the hold of the dump ship Gem *before she sets off to drop them at sea.*

Under Pressure High-powered water hoses are turned on a Greenpeace inflatable as the crew attempts to hold its position beneath the dumping platform.

The exact nature of the material in the drums was never revealed but, whatever the truth of the matter, when the *Gem* set sail for the dumping zone the *Warrior* was six hours behind. Greenpeace planned to position its inflatables under the tipping platforms of the *Gem* to prevent the barrels of waste being dumped off the ship's side.

In fact, Greenpeace arrived hours too late to document the disposal of the two large drums, but the inflatables were soon in position around the *Gem* as her crew began dumping the rest of the cargo, hoisting barrels onto the platform two at a time. As the first of the two drums pitched over the side, a wave swept the nearest inflatable away from the *Gem* momentarily and the barrel crashed into the sea. But the second 270-kilogram (600-pound) barrel smashed onto the dinghy, narrowly missing its two occupants, deflating one of the craft's air panels and destroying the transom and outboard motor. Breaking off the action, the second inflatable took the first in tow and returned to the *Warrior*. The entire incident had been filmed from the bridge of the Greenpeace vessel and was subsequently shown on television around the world.

Having made the first of what was to be a series of actions against the *Gem* in the years to come, the *Warrior* continued south to Spain. Spain was not yet a member of the IWC, and the province of Galicia was the base for Industria Ballenara SA, a company operating four whalers that were killing endangered species and selling the meat to Japan.

By early August the *Warrior* had tracked the whalers, which operated from two whaling stations south of La Coruña, to their hunting ground. As the catcher ship readied its harpoon, Greenpeace launched two inflatables which swept into the line of fire. Despite the arrival of two corvettes from the Spanish Navy, which attempted to stop the protest, the *Rainbow Warrior* managed to confront a second whaling ship.

When the *Warrior* put in at La Coruña, after three weeks at sea, the Navy sent a written request for her to come alongside in the military harbour – tantamount to an order to stay in port. The *Warrior* "declined the offer", shut off her lights when the pilot boat disappeared behind a breakwater, weighed anchor, and quickly left for Portugal.

In October 1978 the *Warrior* was doing battle again, in the frigid waters off the Orkney Islands, north of Scotland. This time she was defending grey seals. They had been on Britain's list of protected species since 1914, but the fishing industry was now claiming that the animals were too numerous and were eating too much of the already depleted whitefish and salmon stocks in the North Atlantic.

To halt this "crisis", the British government called on Norwegian sealers, armed with rifles and hakapiks, to dispatch 900 mother seals and 1,700 pups. Local hunters were licensed to kill 3,200 more pups.

Delphius *The new RI-28, bought by Dutch Greenpeace supporters, gives the sea-borne activists greater speed and range. The fibreglass superstructure provides some protection from wind and waves.*

This was to be the first phase in a six-year culling programme, aimed at reducing the seal population by half, which enraged conservationists throughout Europe.

The *Rainbow Warrior* snapped at the heels of the *Kvitungen*, which carried six expert hunters from Norway to the Orkneys. Behind them were three boatloads of newsmen, with other reporters watching from aeroplanes and helicopters overhead. The unique marine engagement began not far from Scapa Flow, traditional wartime port of the Royal Navy. Wherever the Norwegians went, Greenpeace crews in inflatables darted across their path. Other volunteers on land were poised to chase the seals into the water before they could be killed.

But a major confrontation was averted. The British Prime Minister's office backed down after receiving 16,800 protest letters. It announced it was calling off the Norwegian marksmen, because of "widespread public concern". This was one of the key campaigns that established Greenpeace in Europe.

In June 1979, the *Rainbow Warrior* returned to Iceland, once again to confront the whale hunters. This time she was equipped with a new, highly manoeuvrable 8.5-metre (28-foot) RI-28 inflatable. Financed by Greenpeace supporters in Holland, the RI-28 – named the *Delphius* – was capable of 35 knots. Time and again, whenever a fin whale surfaced to blow out its watery spray, the RI-28 interposed itself between the whaler and the whale. "With growing frustration," reported *Time* magazine, "Captain Thordur Eythorsson stood by his ominous-looking harpoon gun atop the whaler's bow, unable to make his kill."

During her 20-day trip, the *Warrior* was arrested twice by the Icelandic coastguards. An injunction was issued and subpoenas served on the skipper Pete Misson and European director David McTaggart.

The *Warrior* returned in August, and just 10 days after her arrival she was arrested again, following a 24-hour confrontation with two whalers in international waters. Inflatable dinghies worth £20,000 were confiscated by the Icelandic authorities, and it was only after intervention by the British Foreign Office that the equipment was released.

The *Warrior* was arrested again the following year, this time by the French authorities at the port of Cherbourg. The ship's target was the *Pacific Swan*, which, along with her sister ship the *Pacific Fisher*, was being used by British Nuclear Fuels Ltd to bring irradiated fuel elements from Japan to the reprocessing plants at Sellafield (formerly Windscale) in Cumbria, England and at Cap de la Hague in Normandy, France. A report commissioned by Greenpeace from the Political Ecology Research Group concluded that these seaborne shipments represented a major hazard. In the worst circumstances, tens of thousands of lives could be at risk in the event of a fire on board and a consequent

David and Goliath A tiny inflatable harries the Pacific Fisher. *The freighter is carrying irradiated nuclear fuel rods from Japan to France and Britain.*

Making their Mark Greenpeace activists spray their message on the hull of the Pacific Fisher as she arrives in Barrow-in-Furness in Britain. One inflatable was crushed as it tried to prevent the freighter from docking.

1979
Members of Greenpeace Seattle joined in a march to, and occupation of, the Trident nuclear submarine base at Bangor, Washington State. The year before, in a similar action, some 600 people were arrested.

June 2, 1979
Five Greenpeace sky-divers – the "Splat Squad" – parachuted into the construction site of the largest nuclear power plant in the world, the Darlington Generating Station near Bowmanville, Ontario, Canada. The five – Garry Byer, Dan McDermott, Maggie McCaw, Bob Cummings and Don Flint – were all arrested and charged with petty trespass and violations of the Aeronautics Act.

Fifty-eight others, including Greenpeace people from Toronto and Montreal, used the diversion of the sky-dive to scramble over and under the fence into the site, where they were immediately arrested.

The following year, an estimated 1,200 people gathered to protest at the Darlington site. More than 100 scaled the fences, and Greenpeace inflatables enabled five people to gain access to the plant's Lake Ontario shorefront. There were many arrests.

April 3, 1980
Five teams of volunteers from Greenpeace Seattle spent a weekend posting some 3,000 signs warning of the dangers of transporting radioactive waste along the rail route from Bremerton, Washington, to Arco, Idaho. This was in protest at proposals to increase radioactive shipments from the nearby naval shipyard.

release of radiation. Areas within a radius of 50 kilometres (30 miles) of the accident would need to be evacuated to minimize the danger.

Despite an order by the French authorities instructing the *Rainbow Warrior* to remain outside their territorial waters, when she intercepted a radio message from the *Pacific Swan*, the *Warrior* chased the ship towards Cherbourg harbour with a French cruiser in hot pursuit. Two tugs and a minesweeper were guarding the narrow harbour entrance but the skipper, Jon Castle, managed to manoeuvre the *Warrior* past them. One of the tugs turned round and rammed the *Warrior* on the starboard quarter as she neared the dock. Some 50 police armed with rifles, bayonets and grappling hooks seized the boat.

The next day, February 15, the *Warrior* was ordered out of France and banned from returning. She headed for Barrow-in-Furness, the nearest port to Sellafield, where she anchored at the entrance to the harbour channel, ready for the arrival of the *Pacific Fisher*. The British Transport Docks Board had already taken out an injunction in the High Court to prevent Greenpeace interfering with the passage of the *Pacific Fisher* and its cargo. When the ship arrived on March 25, four inflatables with eight of the Greenpeace crew on board followed her. They managed to obstruct the docking, but one inflatable was crushed when it was caught between the stern and rudder of the ship.

As a result of this action Greenpeace was fined £500 and three of its UK directors – David McTaggart, Peter Wilkinson and Allan Thornton – were fined £100 each for disobeying the High Court injunction. The judge, Mr Justice Pain, said it would be inappropriate to jail any of them. "That they are honourable people, I accept. I do not think prison is the place for people like them."

CHAPTER EIGHT

SEA OF TROUBLES

By the last year of the decade, Greenpeace's increasingly diverse campaigns were attracting an ever-growing body of supporters. But success was almost to prove fatal. "There was no systematic growth," says John Frizell, who joined the whale campaign in 1976 and later became executive director of Greenpeace. "The ships would go by a place and an office would appear there, in a place like San Francisco. And then a person would go off from there to, say, Boston, to set up a new office. Meanwhile, Greenpeace Europe was developing in parallel with North America but there wasn't much contact between everybody."

Deadly Waste *An illustration by the French artist Alain Goutal.*

When Robert Hunter had resigned as president in April 1977, Patrick Moore had tried to set up an international board of directors that would control Greenpeace activities (especially fundraising) and that would be headed by himself. A veteran of the Vancouver whaling voyage and indeed of the first Amchitka voyage, Moore had been an increasingly important presence within Greenpeace. It was Moore whose picture holding a seal pup during the 1978 mission to Newfoundland had made all the wire services. He had been arrested for interfering with the hunt and was later acquitted. He had also led a brilliantly successful campaign against the massive use of forest insecticides in British Columbia – Greenpeace's first toxics campaign.

As part of his reorganization plan, Moore had also drawn up a charter vesting ultimate authority in Vancouver but all of the American groups had refused to sign. "Vancouver veterans clung to the notion that every office except the one on Fourth Avenue was a regional office," wrote Robert Hunter. "But San Francisco and the little grass-roots offices scattered about the United States didn't see it that way. It was soon the American Revolution all over again, with the Canadians as the distant colonial overlords."

The founding office in Vancouver desperately needed money: it was at least $140,000 in debt (some put the figure at $250,000). Other groups, particularly that in San Francisco, were increasingly successful at raising funds, and Vancouver felt they were taking advantage of the Greenpeace name – the name that Vancouver had done so much to build up. The climax came when Greenpeace Vancouver sued the office in San Francisco for violation of trademark agreements. San Francisco retaliated by suing for slander.

Juggling Act A cartoon by Terry Peters illustrated the difficulties that the various Greenpeace offices faced as the organization grew and factionalism increased.

Gotcha! This picture of Pat Moore being arrested by the Canadian authorities was reproduced around the world and became a key image in the seal campaign.

DOLPHIN MASSACRE

Three notable actions by Greenpeace activists in Japan, in protest
against the killing of whales and
dolphins, hit the headlines between
1979 and 1981.

In 1978, US environmentalist
Dexter Cate first witnessed the destruction of over 1,300 dolphins by
fishermen on the Izu Peninsula, 200
kilometres (120 miles) southwest of
Tokyo. It was a sight he was unlikely to forget.

He arrived to find some 300 dolphins penned in nets at the harbour
mouth. The water in the harbour
was red with blood, following an

*"Free Dexter Cate" The Hawaiian
branch of Greenpeace campaigned
for Cate's release from prison.*

earlier round-up. As he watched, a
fisherman in one of the boats surrounding the nets hurled a spear
into the milling dolphins, piercing
one of them. This random wound-
ing served to keep the dolphins in a
state of panic and confusion, and
prevented them from escaping
since they will not abandon an injured member of their group.

A dozen dolphins at a time were
then winched ashore by their tails.
What followed was horrific, as the
fishermen went to work with their
knives. "Dolphins with bellies slit
open thrashed about, whistling in
distress as their entrails flopped on
the concrete. They didn't lose con-
sciousness even as blood gushed
from their throats. As we stood
there, horrified witnesses, a fisherman deftly severed the heart from a
still quivering dolphin and tossed it
aside. It landed less than a yard
from my feet, still beating."

The corpses were later dragged to
a mincing machine and the flesh
was ground up for use as pig food
and plant fertilizer.

Cate learned that the dolphins were
being held responsible for the
alarming decline of yellowtail fish
and squid, but he was certain that
pollution and overfishing were to
blame. When he returned to Japan
late in 1979, he discovered that the
government now encouraged the

*Still Smiling Satisfied at the success
of his action, a handcuffed Patrick
Wall arrives at court in Japan.*

Blood Bath *Trapped in the Japanese slaughter pens, dozens of dolphins are hacked to death. The shocking sight of this massacre spurred Dexter Cate and Patrick Wall into action.*

killing and were paying a bounty of $80 per head. It was then that Cate turned to civil disobedience.

In the early hours of February 28, 1980, Cate paddled a small inflatable boat out to Tatsunoshima (Dragon Island), an islet in the bay 800 metres (2,600 feet) off Iki Island, where he untied three ropes and cut a fourth to release 300 dolphins.

When the fishermen returned to their nets, Cate recalls: "They were angry, but not abusive. They understood, finally, that I had acted from a moral position. They just didn't understand that position."

Neither did a Japanese court. Following his arrest, Cate was charged with criminal damage and denied bail. He spent three months in jail before being given a suspended six-month sentence and deported to his home in Hawaii.

Immovable *Patty Hutchinson, flanked by angry crew members, sits chained to the harpoon of the whaler Ryuko Maru.*

Greenpeace member Patrick Wall was shocked by what he saw when he travelled to Japan in November 1980 to research the dolphin slaughter at Izu, south of Tokyo. Here dolphins were killed for food every week during the season.

In January 1981, he managed to dismantle 15 metres (50 feet) of the net barricading the slaughter pens and release 150 dolphins before

daybreak. Two days later he gave himself up to the authorities; after three trials and 62 days in prison, he was given a suspended six-month sentence, a three-year probation and told that he couldn't return to Japan for one year.

In March 1981 it was Patty Hutchinson, a 23-year-old Greenpeace activist from Ohio, who made the headlines when she boarded the Japanese whaler *Ryuko Maru* from an inflatable off the coast at Wadaura Port, Japan. She distributed leaflets to the crew of the whaler, blaming whaling companies and the Japanese government for "abuse of the sperm whales", and then chained herself to the harpoon with two locks, throwing the keys into the sea.

Her protest against the killing of sperm whales forced the *Ryuko Maru* to return to port and prevented her returning to the whaling grounds at the height of the season. Patty Hutchinson was arrested but released later that day without being charged.

Save the World *This UK-produced badge offers the hope that Greenpeace can act as a lifebelt to rescue the planet.*

Show of Force *A Spanish naval corvette arrives to intercept the Rainbow Warrior as Greenpeace inflatables manoeuvre between the whaler* Ibsa III *and its prey.*

On the eve of the court case David McTaggart swept into America calling for unity. He brought together the offices in the United States and Toronto, along with the closely knit groups of Greenpeace Europe, which he headed, and they confronted the old-timers in Vancouver.

The outcome was that Greenpeace Europe paid off the Canadian debts and the US, Canadian and European groups agreed to the formation of a new umbrella organization – Greenpeace International. Its headquarters would be in the Netherlands, where it was registered as a charity under the name Stichting Greenpeace Council (a "stichting" is a Dutch form of organization often used for non-profit-making charities). McTaggart would stay on as the chief executive officer and chairman.

Although there were some who saw McTaggart's action as a personal power play, there was no doubt that it was he who invented the mechanism that rescued the organization from utter chaos. "A lot of the Greenpeace concepts came from Hunter," sums up Jon Hinck, who watched the action from a ringside seat in Seattle. "But it took McTaggart to weave it together."

Under the new plan, each national group would maintain its autonomy to a large extent and devise its own local campaigns. An overseeing council with representatives from each of the member countries would meet to make the major decisions. Greenpeace was now composed of groups in Canada, Australia, the United Kingdom, France, the Netherlands, New Zealand and the United States (which had nine regional offices). Eventually, each voting country would contribute about a quarter of its income to Greenpeace International. Later a five-member board was elected – two from Europe, two from North America and the South Pacific nations, and McTaggart himself.

Such international coordination was becoming increasingly necessary. By early 1980 there were 25,000 paying supporters in the Boston area alone. In the Netherlands new members were being signed on at a rate of 1,100 per month.

A key issue in the Netherlands was toxic waste. The Greenpeace office in Amsterdam had learned that the chemical company Bayer AG in Leverkusen, Germany, was dumping an estimated 550,000 tonnes of acid waste into the North Sea every year. The waste, containing heavy metals and cancer-causing organo-chlorine compounds, was transported down the Rhine in barges and then pumped onto two ships that left for sea from the harbour at Rotterdam. Greenpeace decided it would draw as much public attention as possible to the issue.

On April 25, 1980, a Greenpeace worker attempted to photograph the ships at the dockside and was manhandled by Bayer security guards. Then, on May 20, the *Rainbow Warrior* sailed into Rotterdam to stop the two ships from leaving the port, and Greenpeace members hung banners saying "Bayer Thinks of Today – Tomorrow", mocking the chemical company's advertising slogan. The *Warrior* was accompanied by five other vessels and together they formed a blockade: for three days Greenpeace stopped the ships from dumping 10,000 tonnes of acid waste. By the time the blockade was called off, the dumping was front-page news and the subject of wide public protest. Within two years Bayer decided to halt the dumping of such waste at sea.

On June 6, 1980, Greenpeace's famous flagship left the Port of London, heading once more for Spain and a potentially hostile reception. Remi Parmentier, a director of Greenpeace France, had already been arrested by the Spanish authorities, accused of spying on a whale factory in May, and had been physically threatened.

With reporters from the two largest magazines in Spain and a British television news crew on board, the *Rainbow Warrior* stationed herself off the coast of Vigo. The plan was to confront the whalers using the long-range RI-28, but it was found to be damaged and could not be repaired on the spot. As a result the *Warrior* had to place herself in the middle of the action, and use the smaller Zodiacs.

IN PURSUIT OF THE WHALERS

After a fruitless search up the coast to Corcubion, the site of one of the whaling stations, the *Warrior* headed back towards Vigo, where her crew spotted a whaler, the *Ibsa III*. Almost immediately the Zodiac inflatables were in the water and manoeuvring about the catcher boat, making it difficult for the harpooners to hit their targets.

Greenpeace was so effective at interrupting the hunt that a Spanish corvette, the *Cardaso*, arrived to halt the protest, soon to be joined by the *Pinzon*, which had orders to place the *Warrior* under arrest.

The radio room was sealed off and, under armed guard, the *Warrior* was forced to the naval base at El Ferrol. Determined that this time the *Warrior* would not escape, the Spanish authorities mounted a 24-hour guard and removed an essential component from the propulsion system – the top half of the thrust block, which held the propeller shaft in place. Skipper Jon Castle was taken into custody and later charged with interfering with Spanish fisheries.

For five months the *Warrior* remained in custody. The crew were allowed to go ashore as long as they stayed within the city limits, and they used the opportunity to gain public support through meetings and slide shows. Banners were strung on the arrested ship, saying, "Libertad Para Del *Rainbow Warrior*" (Free The *Rainbow Warrior*). Visitors flocked through the ship's exhibition room.

Eventually parts for the damaged RI-28 speedboat arrived. Having repaired the boat, Bruce Crammond, Chris Robinson and Athel von Koettlitz took off in it across the Bay of Biscay. After a harrowing 800-kilometre (500-mile) ride, they reached Jersey in the Channel Islands, but engine trouble prevented them from making it to the 1980 IWC

Under Arrest *An armed Spanish guard, seen through a porthole of the disabled* Rainbow Warrior, *patrols the quayside at El Ferrol.*

Missing Link *The new thrust block, smuggled in from England, is safely aboard the* Rainbow Warrior *and ready to be fitted.*

Free at Last *The crew of the* Rainbow Warrior *– (from left) Pierre Gleizes, Tony Marriner, Athel von Koettlitz, David McTaggart, Jon Castle, Tim Mark and Chris Robinson – are relieved to reach the island of Jersey after their daring escape from the Spanish military.*

meeting in Brighton on the south coast of England, where they had hoped to tell the world about Spain's illicit whaling activities.

Over the years, Greenpeace lobbyists had been making headway with the IWC. No longer was it simply a front for the whalers. At this meeting the commission cut the whale quotas by 13 per cent and brought orcas under the quota system for the first time. The first steps were taken towards banning the use of the cold harpoon (one that carries no explosive, often maiming rather than killing), on all whales except the minke. The whalers' argument was that an explosive harpoon destroyed too much of the meat of this small whale. The voting failed by only a narrow margin to bring into effect a 10-year moratorium on commercial whaling .

Much of this was good news for Greenpeace, but if the impetus of the anti-whaling campaign was to be kept up, the *Rainbow Warrior* would have to be retrieved without delay. Most of her crew had by now left Spain, although Jon Castle had been forced to remain behind. The Spanish authorities were demanding that Greenpeace pay 10 million pesetas to compensate the whaling company for estimated losses before

they would release the vessel; although Greenpeace had lawyers working on the case through the Spanish courts, progress was painfully slow.

Unable to find a second-hand thrust block for the *Warrior*, Greenpeace obtained the ship's original blueprints and a small engineering firm in England made the crucial part for them. Tony Marriner and the *Warrior*'s chief engineer, Tim Mark, drove down to El Ferrol with the 68-kilogram (150-pound) block and managed to smuggle it past the guards on a dark night while the crew created a diversion.

Very little adjustment was needed to install the vital part but the big question was, would it work? A surreptitious trial of the main motor went smoothly but the real test would come only when the boat was underway. One more problem had to be solved. Five months at anchor meant that the hull of the *Warrior* was encrusted with a heavy growth of weed and barnacles, which were clogging the sea-water intake and causing the engines to overheat. Forty-five minutes' underwater working with a scraper was the solution.

THE WHALING PIRATES

The earliest and most notorious pirate whaling fleet was that owned and directed by Aristotle Onassis, whose activities were reported to the IWC in 1955.

But it was at the 1979 IWC meeting that detailed revelations about the pirate whaling industry really hit the headlines. The exposé, the result of arduous and dangerous undercover work by members of Greenpeace and other conservation groups, revealed that some members of the IWC – principally Japan and Norway – were buying whale products from non-IWC members. Chile, Brazil, South Korea, the Philippines, Spain and Taiwan were all whaling outside IWC regulations with Japanese capital, vessels, equipment and expertise.

The most controversial pirate whaler of recent times was the *Sierra*, owned by Norwegians and operated by South Africans with Japanese technicians on board; the meat was sold to Japan and the whale oil to Norway and the EEC.

Inside Information

The *Sierra*'s activities came to the attention of Greenpeace when the organization received film footage, taken by a crew member on board the ship, showing whaling in progress and grim scenes of a dying mother whale being supported by her distressed calf. The scale of the whaler's operations and her international connections were exposed initially by Nick Carter, a freelance conservationist, and later by Remi Parmentier, Chris Robinson and David McTaggart who, with a British Thames TV team, made the first film of the captain and crew at

Stop Japanese Whaling The pirates' skull and cross-bones on a Japanese flag conveys the campaign message.

work. The *Sierra* was killing every whale that she could find, including juveniles, nursing mothers and such endangered species as the humpback, right and blue. Her career was brought to an end on July 16, 1979 when Paul Watson's boat *Sea Shepherd*, which had been reinforced with concrete, rammed her twice. The *Sierra* limped back to Lisbon for repairs that would take six months. On February 6, 1980 she was sunk in Lisbon harbour by a limpet mine, supposedly placed by Spanish extremists.

On April 27 that year, two Spanish whaling ships, *Ibsa I* and *Ibsa II*, that had also been involved in illegal whaling, were sunk by explosives in the port of Vigo. No one has claimed responsibility or been charged with these actions but again Spanish extremists were believed to be the culprits.

Greenpeace, with its long history of non-violent direct action, worked

in other ways to defeat the pirate whalers. In 1979, the group began investigating the activities of the Taiwanese pirate whaling fleet, which was using four ex-Japanese fishing vessels, converted and operated with finance from Taiyo, Japan's largest fisheries corporation.

The research showed that, despite all her denials, Taiwan was taking an annual catch of 1,000 whales, mostly Bryde's, in a relatively limited area, amounting to the IWC quota for the entire Pacific. The meat was being exported to South Korea, an IWC member, where it was "laundered", repacked and shipped to Japan.

The Scandal Exposed

Greenpeace turned over the results of its exhaustive investigations to the *Sunday Times*, which published a major article on the scandal two weeks before the 1980 IWC meeting. Japan was highly embarrassed by this proof of its continued violation of IWC regulations and banned further shipments of Taiwanese whale meat from entering the country. Greenpeace USA then lobbied through the State Department and sympathetic senators to bring diplomatic pressure to bear on the Taiwanese. As a result the four whaling ships were decommissioned in 1981.

Scuttled in Harbour One of two Spanish whaling ships sunk by unknown saboteurs rests on the seabed at its mooring.

On the moonless night of November 8, 1980, after a strategy meeting in a local bar with David McTaggart, who had arrived secretly for the venture, the *Warrior*'s seven-man crew were ready to make their move. When the guards unexpectedly left their post, the *Warrior* quietly slipped her moorings and eased away from the dock, heading for the harbour entrance and the open sea.

After many tense hours spent listening for the warships and helicopters that were in pursuit – as it turned out they had miscalculated the *Warrior*'s position – the Greenpeace flagship made it to freedom and a rousing welcome in Jersey. Back in Spain, an admiral was dismissed for allowing the boat to escape, and the affair gained Greenpeace even greater support from the Spanish public and led to the establishment of a Greenpeace office in Spain.

The Sirius *This impressive new campaign vessel was to establish its own legend as the Greenpeace flagship in Europe.*

The *Warrior* was now a celebrity. Soon she would prepare for her first trip across the Atlantic, to join in the protest against seal hunting that was planned for the spring of 1981.

Her place in Europe was taken by the *Sirius*, sold to Greenpeace Netherlands by a sympathetic Dutch government for a mere 20,000 guilders, the purchase being further aided by a large donation from the Dutch branch of the World Wildlife Fund.

Built in the Netherlands in 1950 as one of seven pilot vessels for the Royal Dutch Navy, the *Sirius* is 46 metres (152 feet) long, weighs 440 tonnes and has a maximum speed of 13 knots, a maximum sailing range of 30 days and accommodation for 32. Following her purchase, she had to be made ready for her first campaign in just 10 weeks.

By now Greenpeace was firmly established not only in the Netherlands but also in Denmark, where operations began in Copenhagen during 1980, helped by a $5,000 grant from Greenpeace International.

That year, eight protesters from Denmark and the Netherlands jointly staged an action against Norwegian involvement in sealing. Norway was the main exploiter of hooded seals in the North Atlantic, taking 85 per cent of the 1½ million hooded seals killed since 1945.

The activists boarded the sealer *Kvitungen* in Ålesund, chained themselves to the ship and hoisted banners. Photographs of the protest were widely published, and generated a lot of support for the Danish group, whose membership grew to 20,000 over the next two years.

Greenpeace support was also growing in West Germany, where the first office opened in Hamburg in February 1981. One of the early members was Gerd Leipold, an oceanographer and physicist from the Max Planck Institute. "I just felt that being a scientist wasn't enough to change things," he says. "The time was right for an activist group."

Another early member was Monika Griefahn, who had worked for the YMCA in an adult education programme and had been following the Greenpeace whaling campaigns since 1978. In 1980 she had met Remi Parmentier and David McTaggart in France and, with their help, Greenpeace Germany was established.

Within a few months, they had gained more than 3,000 supporters, attracted by such daring actions as that by two Greenpeace climbers on June 24, 1981. They had scaled a polluting smokestack at a pesticide factory in Hamburg and spent the night there, hanging a banner that said, "When the last tree is cut and the last fish killed, the last river poisoned, then you will see that you can't eat money."

This was powerful stuff to the Germans, whose country is heavily industrialized and whose forests were withering under the assaults of acid rain. Germans were especially upset by reports that toxic chemicals were showing up in water supplies and in women's breast milk.

Designing and engineering the German actions was Harald Zindler, a friend of Griefahn's who had been active in the anti-nuclear protests of the 1970s. He had also worked with fishermen who were worried about the quality of the River Elbe, where the fish were growing ugly tumours as a result of pollution. In an early action, deformed fish were dumped outside the German Hydrographic Institute, the government body supposedly responsible for controlling water pollution. It was the issue of chemical contamination that would make the West German Greenpeace office the most active in all of Europe.

A GAME OF CAT AND MOUSE

On March 19, 1981 the *Rainbow Warrior* arrived off the east coast of Canada to play her part in the protest against the killing of 15,000 hooded seal pups. Six days later, after playing a game of cat and mouse with a Canadian fisheries protection vessel, the MV *Baffin*, she was towed unceremoniously from the area, and campaign organizer Allan Thornton was arrested. Crew members Chris Robinson and Willem Beekman, of the Netherlands, were also arrested and charged with violation of the Seal Protection regulations because they had sprayed the pups' fur with green dye to make the pelts commercially valueless. A fine of $2,000 and jail sentences suspended for three years were imposed on the three Greenpeace protesters. Nonetheless, the Greenpeace presence had forced the sealers to cut short their slaughter.

***At the Helm** Monika Griefahn, seen here on the bridge of the* Sirius, *played a key role in the growth of Greenpeace in West Germany. She became an international board member in 1983.*

Campaign Voyage *A cheerful crew* (right) *gathers on the deck of the* Rainbow Warrior *as she leaves the docks at Southampton, England, at the start of her first Atlantic crossing in the spring of 1981. On reaching the east coast of Canada, the* Warrior *picks her way carefully through the pack ice* (below) *to the sealing grounds, where the crew will attempt to disrupt the planned slaughter of 15,000 hooded seal pups.*

Saving the Seals Campaign
organizer Allan Thornton (top
left), *Willem Beekman and Chris
Robinson* (left) *were arrested by
fisheries officers and Royal
Canadian Mounted Police. A trail
of blood on the ice* (top right) *is
the cruel reality behind these
expensive fur coats* (above).

In Dry Dock *Greenpeace members save time and money by taking care of the* Rainbow Warrior's *refit themselves.*

After her release, the *Warrior* moved on to Boston, and this gave the office there an enormous boost. The canvassers going from door to door for funds throughout New England were imbued with fresh enthusiasm now they had a tangible source of inspiration.

But the Atlantic crossing had taken its toll on the ship and she was badly in need of repair. The *Warrior*'s crew "had gotten ready to go very quickly, and had had a rough crossing," recalls Steve Sawyer, at that time working for Greenpeace International as the ship's manager. "They bent the propeller, everything leaked, the electronics didn't work, the wiring was all screwed up, the motors were not in good shape, the scuppers were clogged, and they had sailed across the Atlantic with about six inches of water inside the bridge. It was a total mess."

It was Sawyer's job to get her back in shape. On the way back from a trip to Georges Bank, a productive fishing area off the northeastern coast of the US, where the *Warrior* had protested against plans for oil drilling, fire broke out in the engine room and part of the propulsion system was ruined. Faced with an estimate of six months and $650,000 to have the ship repaired professionally, Sawyer and his group took the *Warrior* to Stonington, Maine, and did the job themselves at a considerable saving and in half the time. "At the end we had a boat that was a little faster, and had twice as much range and manoeuvrability," says Sawyer. "In short, a new lease on life."

INTERNATIONAL ACTION

In 1981 Greenpeace was a whirlwind of activity. Protests were taking place around the globe, and in the course of the year there were about 50 different actions in support of an increasing number of campaigns.

Greenpeace activists in Cherbourg occupied cranes to stop them loading and unloading spent nuclear fuel. In Tacoma, Washington, they climbed a 170-metre (560-foot) smokestack to protest against acid rain. In Europe the actions against nuclear dump ships proved especially dramatic. Here the *Sirius* was the star.

The ship had left her berth in Amsterdam on July 13 and had sailed south to confront the *Gem* at the UKAEA dump site 1,000 kilometres

STOP THE SUPERTANKER
When the Canadian Coast Guard announced in early 1981 that they were going to "test" a 189,000 deadweight-tonne supertanker to see if it could navigate the inland waters of Puget Sound without spilling oil on the beaches of the tiny islands that dot it, Greenpeace declared that it would stage a protest. The Coast Guard responded by banning spectators within 1,800 metres (6,000 feet) of the tanker and threatening that anyone who ventured closer would end up in jail.

Determined *Patrick Moore defies the Coast Guard ban and pursues the supertanker navigating Puget Sound.*

When the day of the test came, Patrick Moore and Rex Weyler of Greenpeace, together with two reporters, outmanoeuvred the Coast Guard cutters and sailed into the banned zone in rubber boats. They were arrested but released the next day, when they immediately gave a repeat performance.

When a motion calling on the House of Commons to express its appreciation to Greenpeace was put to the Canadian Parliament, it only narrowly failed to receive unanimous consent.

(600 miles) southwest of Land's End. On this latest protest voyage against the dump, Greenpeace attempted to catch one of the 3-tonne barrels of waste in an empty inflatable in order to take it ashore for analysis. The *Gem*'s crew used grappling hooks to pull one of the rubber boats on board, where it was destroyed. They dropped another barrel on the outboard motor of the launch *Delphius*, effectively immobilizing it. Several people were injured as the crew of the dump ship used high-pressure hoses to try to swamp the inflatables.

Pete Wilkinson told the *Sunday Times*: "The drums were missing us by inches and there's this man on board shouting 'It's your funeral.' It was unbelievable – we thought somebody was going to get killed." Ignoring the dangers, the Greenpeace campaigners managed to delay the dump for six hours and once again succeeded in drawing the attention of the international press to the issue.

SHIFTING THE BALANCE

At around the same time, the whaling issue was coming to a head. Greenpeace sent telegrams to every nation involved in the IWC, pleading on behalf of the defenceless whales. Victory appeared to be in sight. Said McTaggart, "The balance of power at the IWC is shifting dramatically in favour of the whales." But, once again, the IWC did not declare a moratorium, although it did ban the hunting of sperm whales and agreed to a ban on the use of the cold harpoon after the next whaling season. By the end of the year, the fisheries commission of Spain had called for an end to that country's whaling.

Besides the *Rainbow Warrior* and the *Sirius*, the Greenpeace fleet now also included McTaggart's old ketch, *Vega*, which had been repurchased by Greenpeace. On October 30, 1981, she sailed once more for Moruroa, this time setting out to cross the 6,400 kilometres (4,000 miles)

***Close Shave** A Greenpeace crew in an inflatable from the* Sirius *is blasted by the* Gem's *high-pressure hoses as a falling barrel skims the hull of the fragile craft* (top). *Peter Wilkinson* (above, on the left) *initially identified the UKAEA dump site in the Atlantic.*

of ocean from Manzanillo in Mexico. On board were McTaggart, Chris Robinson, Tony Marriner, Lloyd Anderson and Brice Lalonde (a presidential candidate for the French Ecology Party in the spring 1981 elections in which Mitterand was elected).

This voyage, which marked Greenpeace's renewed involvement in the nuclear testing issue, coincided with publication of a report by workers at the test site that had been leaked to the French press. The report complained about poor security arrangements and revealed that several severe accidents had occurred on Moruroa in recent years. It was estimated that, as a result of the tests, the atoll had sunk 1.5 metres (5 feet) – about 2 centimetres (0.8 inches) at each explosion – and that an underwater crack had formed, about 5 metres (16 feet) wide and 8 metres (26 feet) long, that might be allowing dangerous radiation to leak into the ocean.

PEACE PROPOSAL

Off Moruroa, *Vega* encountered her old enemy the *Hippopotame*. The *Vega*'s crew were informed that if they crossed the 19-kilometre (12-mile) territorial limit surrounding Moruroa they would be arrested immediately. Undeterred and determined to persevere with the mission, Brice Lalonde sent a letter to President Mitterand via the commander at Moruroa and the French high commissioner in Papeete calling for France to suspend her atomic testing programme and to take the initiative in proposing a Comprehensive Test Ban Treaty at the forthcoming United Nations session on disarmament in June 1982.

No reply had come back by the fortieth day of the voyage and food and water on board *Vega* were running low. At the last moment the crew heard the French response on the radio via the Greenpeace office in Paris. The government would not stop testing but it would allow an independent scientific survey of Moruroa's flora and fauna. Believing such a survey would reveal unacceptable levels of radiation in the environment, the crew felt they had achieved their purpose and set sail for Tahiti.

However, this independent investigation was not allowed. Instead, in June 1982, the French government sent its own eight-man team, led by vulcanologist Haroun Tazieff, into Moruroa where they took air and water samples during a test to verify safety arrangements. The report published a year later played down most of the risks, described the effects of radiation as "feeble and innocuous" and concluded there was little risk of contamination.

As a result Greenpeace broke off direct communication with the French government. "We felt," said Remi Parmentier, "that France was only pretending to cooperate with a study."

Further protest voyages to Moruroa were to come in the years that followed but, for the moment, Greenpeace turned its attention to another nuclear power – the Soviet Union.

Moruroa Bound The Vega *in full sail* (opposite) *heads away from Manzanillo, Mexico, on her third voyage to the French nuclear test site. She is crewed by* (above, from left to right) *Chris Robinson, Lloyd Anderson, David McTaggart, Tony Marriner and Brice Lalonde.*

CHAPTER NINE

NEW FRONTIERS

THE GREENPEACE SHIP **Sirius** *was chosen to take a crew of 19, representing six different nationalities, together with eight members of the press, into the port of Leningrad, setting out from Amsterdam on May 12, 1982. The action followed a decision by Greenpeace to step up its anti-nuclear activities around the world – and it could not ignore the USSR.*

The Trinity *A hot-air balloon takes up the Greenpeace message.*

The Soviet authorities had conducted 467 of the world's 1,300 nuclear tests to date. Like the United States and France, they exploded their bombs underground, but they provided no information on the release of radiation during the tests. Greenpeace had requested that the USSR allow an independent team of scientists to research the effects of the explosions on the Arctic island of Novaya Zemlya, but they had refused. The time had come for action.

En route to the USSR, the *Sirius*, under captain Willem Beekman, was to make several stops in Scandinavia in an effort to promote the anti-nuclear cause. On May 15, she joined a peace rally in Göteborg, Sweden, and on May 18 she held a press conference in Copenhagen. She arrived in Helsinki a week later. There the crew demonstrated in front of a hotel where a meeting of the Socialist International was being held, and announced that they were leaving for Leningrad on May 31, even though they had no visas.

After discussion with a member of the Soviet Peace Committee, visas were hurriedly issued, but they allowed Greenpeace a visit of only three days. Before the *Sirius* left, she took on board Daniel Ellsberg, the former US Deputy Assistant Secretary of Defense turned anti-war activist, who had leaked the *Pentagon Papers* – the classified history of US involvement in the Vietnam war – to the media in 1971.

AWKWARD GUESTS

By June 2 there had been no word from the ship for 48 hours, but then came news that the *Sirius* had anchored safely at Kronstadt, some 32 kilometres (20 miles) from Leningrad, the day before; the Soviet authorities had been on board to offer the crew a sightseeing tour of the city and a meeting with Soviet officials and reporters from Moscow the following day. However, Greenpeace wanted a press conference to be held on the *Sirius* straight away, and eventually the Soviets agreed.

After the conference, the Soviet authorities proposed a visit to a World War II cemetery: when the Greenpeace crew declined the invitation, they were restricted to the vessel until the following day.

January 16, 1982
Greenpeace activists in California flyposted some 2,000 "Radioactive Alert" posters along the major routes in Northern California used for the transportation of radioactive material.

February 8, 1982
Greenpeace members James Greager and Bob Jacksy reached the top of the 260-metre (850-foot) smokestack of the American Electric Power Company's Conesville coal-burning plant, 100 kilometres (62 miles) east of Columbus, Ohio. At the same time, Jeff Petteren and James Stiles were atop the 200-metre (650-foot) smokestack of the Indiana and Kentucky Electric Company's Clifty Creek plant in Madison, Indiana.

A day later, two more Greenpeace activists – Claire O'Brien and Davis Stewart – started their ascent of the 168-metre (550 foot) stack of the Magma Copper Company's smelting plant in San Manuel, Arizona.

All these protests drew attention to the emission by these plants of sulphur dioxide, a major cause of acid rain.

All the climbers were later charged with criminal trespass.

March 8, 1982
The Sirius *joined eight German fishing vessels in a protest blockade outside the Dow Chemicals plant near Hamburg. All river traffic to the company was halted as a result. The fishermen asked Greenpeace for its support when the West German government issued a ban on the sale of all fish caught in the River Elbe, after some were found to be contaminated with organochlorine chemicals and mercury – toxic substances dumped in the river by Dow.*

Body Language *Greenpeace members – (left to right) Hans Guyt, Bart Romijn, Anna Stoffel, Daniel Ellsberg and Thyra Quensel – hold a press conference on board the* Sirius *in Kronstadt harbour. Their T-shirts bear the campaign message in Cyrillic script.*

April 20, 1982
Greenpeace called upon the Belgian government to introduce a law controlling pollution of the North Sea. Belgium was the only country in the EEC not to have such a law and Greenpeace feared that Belgium would become the main outlet for Europe's industrial waste.

Several hundred MPs, Euro-MPs, trade unions and scientists from 11 European countries supported Greenpeace's concern for the North Sea.

May 10, 1982
Greenpeace Denmark and Greenpeace Germany held a ceremony for the birds that had vanished from the Danish part of the Wadden Sea. The ceremony coincided with the official opening of a new dyke – planned in 1976 as protection after a big flood that year – by the Queen of Denmark and the Federal President. By building the dyke, Denmark had gained a piece of agricultural land but had lost an internationally unique habitat. Since its construction, 85 per cent of the wading birds and 65 per cent of the duck species had migrated and failed to return.

At the meeting with Soviet officials at the House of Friendship, Greenpeace presented a statement of concern over nuclear testing signed by scientists and politicians, and handed over a telegram addressed to President Leonid Brezhnev calling on the Soviet Union to declare a unilateral test freeze. But members of the officially-sanctioned peace movement refused to give Brezhnev the telegram; all the Greenpeace crew could do was hand out leaflets in the streets.

The Soviets had had about enough. On Wednesday evening they asked the *Sirius* to leave. The crew announced that they would not leave until they received a response from Brezhnev, and very shortly they found themselves being towed out of the port by two tugboats - not, however, before they had released 2,000 helium-filled balloons carrying the message in Russian "Soviet Union: Stop The Atomic Tests", despite attempts by Soviet officials to prevent them.

Just two weeks after the Leningrad action, Greenpeace staged a major protest at the US Nevada test site to coincide with demonstrations by other peace groups at the Lawrence Livermore laboratories near Oakland, California, one of two development centres for nuclear weapons in the United States.

Greenpeace launched a hot-air balloon, the *Trinity*, at the edge of Highway 95, near the entrance to the test site. The balloon, tethered 46 metres (150 feet) above the ground and bearing a 6-metre (20-foot) banner with the message "Stop Nuclear Tests", stayed aloft for over two hours, piloted by Gene Stilp and Saul Bloom.

Greenpeace was also doggedly pursuing its campaign to prevent both the international transport of spent nuclear fuel and the dumping of nuclear waste at sea.

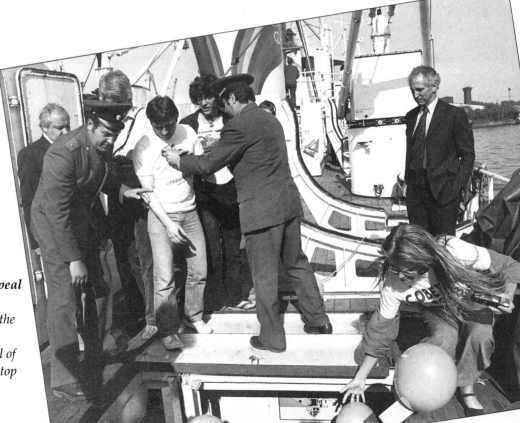

Launching an Appeal
Soviet authorities attempt to prevent the Sirius *crew from releasing a hold-full of balloons saying: "Stop Nuclear Tests".*

THE CEDARLEA

Built in 1962, the *Cedarlea* is an ex-North Sea side-trawler, formerly named the *Welsh Monarch*, which was offered to Greenpeace for a knock-down price of £5,000 in 1981, an offer that was difficult to refuse. The *Cedarlea* is 40 metres (130 feet) long and 308 gross tonnes in weight with a maximum speed of 30 knots and berths for 15.

When she was purchased she was in a very dilapidated state and, after being moved from her home port of Ipswich to the Royal Docks in London, she was temporarily mothballed. Then in the summer of 1982 a frenzied three and a half weeks of conversion by up to 30 volunteers a day brought her up to scratch.

MV Cedarlea *A new ship joins the growing Greenpeace fleet.*

Freshly painted in Greenpeace colours, the *Cedarlea*'s first mission was to sail to Brighton where, along with the *Sirius*, she lay in silent vigil facing the hotel where the IWC meeting was taking place. One of her inflatables ferried members of the public out to look around the new campaign vessel.

When the meeting closed, the *Cedarlea* accompanied the *Sirius* to protest against nuclear waste dumping. The following year she was involved in three campaigns: to expose those responsible for chemical pollution in the River Humber; to monitor radioactive discharges from Sellafield; and, in Belgium, to prevent the dump ship *Falco* from discharging its cargo of toxic waste into the North Sea.

In 1984 she was sold to a Scottish company who converted her for work as a standby vessel serving North Sea oil-rigs.

In a five-week period during August and September of 1982, in the largest operation of its kind, four cargo ships were to dump some 15,000 tonnes of nuclear waste from the UK, Belgium, the Netherlands and Switzerland in the North Atlantic off the northwest coast of Spain. Greenpeace was determined to do all it could to hinder this operation.

The first encounter was between the *Sirius* and the UK dump ship *Gem*. Steel cages had been erected around the four dumping platforms to prevent Greenpeace using its inflatables. So on August 10, in a change of tactics, six Greenpeace members chained themselves to the dumping platforms and occupied them for 77 hours. When the *Gem*'s crew rigged up a fifth platform, the inflatables went into action and were blasted by high-pressure hoses. The UKAEA later admitted that the three days of action by Greenpeace had hindered the dump.

As a result, the Dutch nuclear authorities and the UKAEA resorted to the courts. An injunction was upheld in British courts preventing Greenpeace Netherlands from interfering with the dump, but it was acknowledged to be difficult to enforce. So in September the UKAEA went to the Netherlands and gained a partial victory in the courts there. The court recognized Greenpeace's right to carry out protests at the dump site but not its right to make the dumping impossible or to board the dumping vessel. A fine of £2,000 would be levied for each day the organization failed to comply.

The Dutch nuclear authorities asked the courts to: prevent Greenpeace vessels going within 1 kilometre (1,100 yards) of the dump ships; prevent "supporters and agents" of Greenpeace carrying out actions of protest; make direct actions illegal; order that stopping the dump would cost Greenpeace 60,000 Dutch guilders a day. The judge threw out all these demands except the last, thereby recognizing Greenpeace's right to protest while protecting the "right" of the nuclear authorities to pollute international waters.

Locked Out *A steel cage, erected around the* Gem's *dumping platform, is designed to keep the Greenpeace inflatables at bay.*

July 21, 1982
After hearing evidence from Greenpeace and other environmental groups, the Dutch Council of State announced that two companies, Kronos-Titan and Pigment Chemie, must halve their dumping of titanium dioxide waste.

July 22, 1982
Greenpeace Sweden and two other anti-nuclear groups sent a letter to the Swedish prime minister protesting against further export of Swedish spent fuel and its reprocessing.

Near-Disaster *Gijs Thieme drives his Zephyr inflatable under the dumping platform of the* Rijnborg *just as two barrels are dropped (top). The barrels, which weigh almost a tonne, crash down on the bow of his craft and catapult him into the sea (middle). The second Zephyr moves in quickly and Thieme is pulled from the water, unhurt but badly shaken (bottom). This incident, which could so easily have been a tragedy, did not stop the protest action.*

While these legal machinations were going on, the second phase of the Greenpeace operation was underway with the *Sirius* attempting to prevent the Dutch ship *Scheldeborg* dumping its 3,000-tonne cargo of nuclear waste. Inflatables were repeatedly driven under the dumping platforms until two barrels hit one of the dinghies and tipped its crew member into the water. The crew of the *Scheldeborg* apologized for the incident and stopped dumping for the day.

After refuelling and reprovisioning in Brest, the *Sirius* returned to the dumping site where it joined forces with the *Cedarlea* against two Dutch dump ships, the *Maryke Smits* and the *Rijnborg*, carrying 7,000 tonnes of Swiss and Belgian waste. The *Sirius*, under the captaincy of Willem Beekman, and the *Cedarlea*, under captain Ken Ballard, concentrated their efforts on the smaller *Rijnborg*. The battle between the dump ships and the inflatables was as risky as ever.

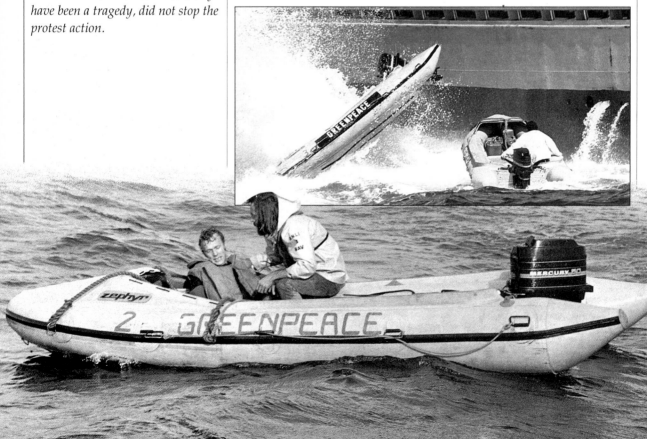

"These barrels weighed about 900 or 1,100 lbs each, and there was a crane that took two, three, or four barrels at the same time and swung them over the side," says Greenpeace member Gijs Thieme. "You drive under the railing and try to look at the person who is dumping - just to look at his face and keep contact, so he won't pull."

But the dumpers did not always hold back and, as Thieme himself discovered, sometimes it was too late to swerve away. As he manoeuvred under the platform, two barrels came hurtling over the side and hit the front of his craft. The inflatable disappeared beneath the waves under the barrels' weight, then catapulted back to the surface – without Thieme, who had to be pulled from the ocean half conscious.

Watching in horror from the *Sirius* was Dutchman Hans Guyt. "I immediately sent telexes to the company, to the press, to the minister of the environment, telling them to pull this dump boat back before people got killed. The rest of the crew got really angry and upset and we decided to stop them once and for all.

"We put Gijs in a big inflatable with some journalists and sent it over to the ship. The entire crew, including the officers, came to the deck to look at this guy who was still alive! That took all their attention away. We had another inflatable on the port side with a makeshift ladder, and our people climbed aboard and occupied the ship's dumping cranes, chaining themselves to them for 48 hours."

The three protesters – American Russell Wray, Harald Zindler from West Germany and Modesto Zola from Spain - prevented the *Rijnborg* from continuing the dump. For 36 hours the ship steamed at full speed on a haphazard course pursued by the Greenpeace boats. The captain arrested the protesters under orders from the Ministry of Justice in the Netherlands but they were subsequently released en route and all charges against them were dropped.

These actions, which were coordinated with other protests in many European countries, led the Dutch government to announce, on September 22, its intention to halt the dumping of nuclear waste at sea and to look for alternative methods of disposal on land.

THE *WARRIOR* IN ACTION

Throughout 1982 the *Rainbow Warrior* was also busy. She now had a new captain, Peter Willcox, a 28-year-old American who had previously skippered an ecology boat patrolling the Hudson River and had sailed on a whale research vessel. In January and February the *Warrior* was in action on the East Coast of the United States, initially in a series of engagements against NL Industries, a multinational corporation that for years had been dumping nearly 4.5 million litres (1 million gallons) of acid-iron waste a day at a site 24 kilometres (15 miles) off the New Jersey coast. The *Warrior's* protests were designed to coincide with an Environmental Protection Agency (EPA) hearing in New York to review NL's dumping permit.

The *Warrior* anchored at the mouth of the Raritan River, and when the crew spotted the huge NL waste barge the *William N. Taft*, four Zodiacs sped across to intercept it, holding red flares aloft. The next morning, when the barge returned, four inflatables again accompanied it.

ONCE MORE TO MORUROA

Greenpeace made its fourth trip to Moruroa in October 1982, just days after the French government announced plans to strengthen its nuclear strike force. Four nuclear tests had already been carried out on the atoll that year, and the last one had been the largest in two years.

The trip was also in protest at the inadequacy of the June 1982 scientific mission to Moruroa, headed by Haroun Tazieff, which had concluded that radiation from the tests posed no risk. Greenpeace and others considered that this conclusion ignored overwhelming evidence to the contrary.

The crew of the *Vega* – Tahitian Guy Taero, the Australian Chris Robinson, and Jon Castle from England – sailed from Papeete in Tahiti and arrived at the 19-kilometre (12-mile) international limit off Moruroa on October 24.

Seven days later, at midnight, eight armed guards from the *Hippopotame* boarded the *Vega* and towed her into Moruroa, where the crew were kept in custody for 24 hours. Castle and Robinson were deported after being forced to sign an agreement never to return to French Polynesia. The *Vega* was later towed to Tahiti by the French.

Intrepid Trio *On the fourth protest voyage to Moruroa, the Vega is crewed by Chris Robinson (left), Guy Taero and Jon Castle.*

SUCCESS FOR SEALS

The seventh year of the Greenpeace harp seal campaign, under the direction of Patrick Moore, was among the most dangerous and rewarding it had ever staged.

The *Rainbow Warrior* arrived in Halifax, Nova Scotia on February 23, 1982 and, the following day, a press conference was held on board featuring not only a Greenpeace spokesman but also representatives from the Fund For Animals, the Animal Protection Institute and the International Fund for Animal Welfare (IFAW). Greenpeace revealed its "secret weapon" for this year's protest – three hovercraft that could travel on solid ice or open water. The two 2-passenger hovercraft and a larger 8-passenger craft had been transported by road from Vancouver in a gruelling eight-day, 6,400-kilometre (4,000-mile) journey through a series of avalanches and in freezing temperatures.

The next day the harassment began. Customs officials seized the *Warrior* until a $400 fine for "improper customs clearance" was paid, a violation the *Warrior* allegedly committed during her visit to the Labrador coast in 1981. A few days later the largest hovercraft was vandalized, then it was damaged beyond repair when the engine cowling blew into the rotor blades.

The heavy ice conditions that year meant the start of the hunt was postponed and all but the two largest sealing ships – the *Brandal* and the *Techno Venture* – were unable to reach the seal nursing grounds.

A Blow to Sealing

The *Warrior* herself arrived on March 10, after a slow and dangerous eight-day journey through the ice, guided by Greenpeace helicopters. The Coast Guard ice breaker *Tupper* and its attendant helicopters provided a sinister escort.

The crew were heartened though by the news on March 11 that the European Parliament had voted by 160 to 10 to recommend a ban on the import of seal skins and products into the EEC. As the EEC comprised 75 per cent of the market for harp seal products, the prospect of such a ban was seen by many as a body blow to the sealing industry.

Direct action started the next day when four Greenpeace members set out for the sealing grounds. One of the remaining hovercraft had engine problems and the other had been damaged in a collision with a large chunk of rafted ice, so they made their journey by foot.

Determined Action

They found the sealers at work, as one eyewitness put it, "turning the pristine seal nursery into a spider-web pattern of red-streaked ice, dotted with frozen carcasses and stacks of seal pelts".

After spraying the pelts of over 200 seal pups with indelible green dye, three of the four were arrested by Royal Canadian Mounted Police and fisheries officials.

The three – Patrick Wall, Greta Cowan and Jos Van Heumen – were charged under Section 20 of the Seal Protection regulations, which prohibits the dyeing of a seal pelt. They spent three days in prison before a court appearance in which they pleaded guilty and were sentenced to fines of $1,500 (Can) each or two months in jail. They were also required to sign a form stating that they would not come to Atlantic Canada during the seal hunt for three years.

Signs of a Breakthrough

By the end of 1982 the tide was turning in the seals' favour. For the first time, major Canadian newspapers were acknowledging that the seal slaughter was more of a liability than a necessity for Canada. Wholesale prices for seal pelts had dropped dramatically in response to the European Parliament recommendation and, on December 17, the EEC import ban was finally introduced. Greenpeace's seal campaign for the year had been a great success and it looked as though the long struggle to stop the annual hunt would soon be over.

Better Dyed than Dead A protester sprays a seal with harmless dye to render its pelt valueless and save it from the sealers.

KANGAROO SLAUGHTER

A new Greenpeace campaign aimed at abolishing the commercial trade in kangaroo products was launched in October 1982. Of the 48 species of kangaroos and wallabies originally found in Australia, seven are now extinct and 12 are considered endangered. Five species are killed commercially for internal use but the main target of the hunt is the five species that are killed for the commercial export industry – the red, the eastern and western grey and the euro and whiptail wallaby. The first three of these species are listed as threatened under the US Endangered Species Act but are not listed as endangered or vulnerable by the Australian authorities. The annual kill of 2–3 million animals is one of the largest slaughters of wild mammals in the world.

The Scapegoat

It is claimed that kangaroos cause extensive agricultural damage and compete with livestock for scarce grazing, but scientific research to date substantiates neither of these claims. In fact the commercial killing zone is primarily restricted to the semi-arid regions of west, south and east Australia, where only around 10 per cent of the continent's sheep and cattle are located.

Greenpeace opposes the hunt on ecological and ethical grounds. In its view the kangaroo is being used as a

scapegoat for decreasing farm productivity, which is actually due to inappropriate land use – more than 50 per cent of Australia's agricultural land now requires treatment for soil erosion.

The authorities in charge of the hunt claim they run a scientifically based cull. Yet the quotas they set, usually 15 per cent of the estimated population, are based on guesstimates derived from aerial and ground surveys. They do not take into account the effect of natural disasters such as the 1982/83 drought – which killed nearly half the population of the three major commercial species of kangaroo in eastern Australia – nor the illegal hunt, which accounts for in the region of one million animals per year. It is economic gain, not agricultural damage, that drives the slaughter.

The killing is indiscriminate and the methods used are inhumane.

Raw Hide *Carcasses hang from the side of a hunter's pick-up truck as he lines up another kangaroo in the sights of his rifle.*

Ninety per cent of the 2,500–3,000 shooters – who hunt at night using searchlights and high-powered rifles – are part-timers who are not trained for proficiency. Large numbers of joeys (baby kangaroos) are clubbed to death during the hunt.

Kangaroos are killed for both meat and hides. Skins are used for fur and leather products; the meat for human consumption and pet food. Over 90 per cent of the skins are used by Italy, Japan and the United States.

The Greenpeace aim of ending the slaughter by stopping the trade met with some success when a poster campaign against kangaroo-skin running shoes led to manufacturers withdrawing them from sale.

After this second day of protest, NL attempted to get a temporary restraining order through the courts, claiming that Greenpeace was part of a "conspiracy… to harass, interfere with business, and destroy the reputation" of NL. This restraining order was dismissed but, meanwhile, NL had reached an out-of-court settlement with the EPA, allowing them to continue dumping. The protest resumed, and this time three Greenpeace members handcuffed themselves to the barge's anchor chain. With its dumping operation now firmly in the spotlight, NL would be under much greater scrutiny in the future.

After playing her part in the protest against the annual seal cull in Newfoundland, the *Warrior* called at St John, New Brunswick, to lend support to widespread opposition to Canadian plans to ship 3,000 nuclear fuel rods to Argentina, intended for the commercial reactor at

July 23, 1982
Greenpeace USA staged a brief ceremony at the Department of the Interior in Washington, at which they presented 5,000 marbles to Interior Secretary James Watt. His decision to open the entire US coastline to industries drilling for oil and gas suggested he had lost them.

"Victory – First One to the Whales" Anti-whaling protesters outside the IWC meeting in Brighton celebrate the commission's vote to end commercial whaling.

July 30, 1982
A Sydney magistrate dismissed charges of trespass against seven members of Greenpeace Australia and told them: "I am a bit for the whales myself." The seven had gone to demonstrate at the Japanese consulate, suspecting that Japan was going to renege on the whaling moratorium.

August 11, 1982
Greenpeace representatives participated in a conference organized in Hiroshima, Japan, by the anti-nuclear group Gensuiken and held in parallel with the 39th World Conference Against the A- and H-Bombs.

Success Hans Guyt (left), John Frizell and Remi Parmentier are delighted by the IWC vote.

Embalse. Greenpeace and others were concerned that this material would be diverted to military ends, as Argentina had refused to sign the Nuclear Proliferation Treaty.

These and similar actions kept the *Rainbow Warrior* busy until later in the year, when she headed off for her first trip to the Pacific to carry out a series of important protest actions against whalers. In the meantime, momentous events took place in Europe that were to make her trip even more timely.

On July 23, 1982, the IWC meeting in England, had voted by 25 to 7 with 5 abstentions to end all commercial whaling in three years' time, an historic achievement for the whale conservation lobby. But, by November 4, four of the eight countries actively involved in whaling had filed objections to the IWC decision. These nations were the Soviet Union, Japan, Norway and Peru. Greenpeace determined to persuade them to withdraw their objections and, because the *Warrior* was by this time relatively near to Peru, that country became the first target.

UNDERCOVER OPERATION

Within three days, the *Warrior* had covered the 800 kilometres (500 miles) to Peruvian waters and had begun searching for the whaling ships owned by the Peruvian Victoria Del Mar Company (Vicmar), which is in fact managed and controlled by the Japanese. After three days the *Warrior*'s crew sighted the whaler *Victoria 7* heading for the port of Paita with a small whale in tow, and followed it into harbour.

On land a furious debate was underway following Peru's IWC objection. The Peruvian Fishermen's Union voted in favour of a whaling moratorium. Vicmar workers and thousands of Paita residents signed petitions against poor working conditions and rates of pay at Vicmar and in support of Greenpeace. Vicmar responded by claiming Greenpeace was paid by a US whaling company. The government was feeling the pressure of world opinion as protest telegrams flooded in.

The authorities reacted swiftly to Greenpeace's presence, fining it $3,000 for "unauthorized use of Zodiacs" after Athel von Koettlitz and Raphael Demandre chained their inflatable to the *Victoria 7*'s stern rail, although the charges were later dropped.

Undeterred, six other members of Greenpeace boarded the *Victoria 7* on December 13 and chained themselves to the harpoon-gun, while the *Warrior*'s captain, Peter Willcox, climbed the whaler's mast and occupied the crow's nest, hanging a banner from it.

As recounted later by Campbell Plowden in the *Greenpeace Examiner*, "One very hot day passed, and as night came on we took turns sleeping by the decommissioned harpoon. Our silent vigil was abruptly ended at 3 a.m. when machine-gun-wielding marines suddenly boarded the vessel. We were detained for the day in the port captain's office, then released into the custody of more guards on the *Rainbow Warrior*, which had been seized simultaneously with our arrest."

For a while the situation looked grim. Following the Spanish example, the Peruvian authorities removed the *Warrior*'s thrust block, sealed off the radio room and placed five armed guards on board. The Greenpeace crew learned that they were in danger of being charged with piracy, which carried a maximum prison term of 20 years.

Fortunately, on December 17, the prosecutor decided there was insufficient evidence to proceed with the piracy charge, and the Greenpeace protesters were released with a $3,000 fine. Two weeks later the *Warrior* was also freed.

Six months later the *Warrior* was poised for another whaling action – one that, in terms of publicity, exceeded even the attention given to the first Pacific whaling voyages of 1975 and 1976. Again, Greenpeace was off to the Soviet Union – this time, Siberia.

The aim of the mission was to film the Lorino whaling station, situated on the Chukchi Peninsula, across the Bering Strait from Alaska. Under IWC regulations, the Soviets were allowed to catch rare grey whales, which are otherwise totally protected, provided the hunt was by aboriginal whaling techniques, carried out by local people as part of their tradition to help them subsist in the harsh environment. But Greenpeace suspected that the whales were being killed merely to provide cheap food for mink being bred for the fur trade.

Timed to coincide with the 1983 meeting of the IWC in Brighton, England, the mission was launched early on the morning of Sunday, July 17, from Nome, Alaska. Twenty-four hours later the *Warrior* turned directly into Soviet waters towards the whaling station. They had now entered the most forbidding region of the Soviet Union – without permission.

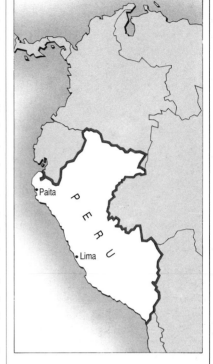

In Peru The Warrior *goes to Paita in an attempt to alter the government's attitude to whaling.*

Passive Resistance *Chained to the harpoon, Campbell Plowden helps to put the Peruvian whaler out of action.*

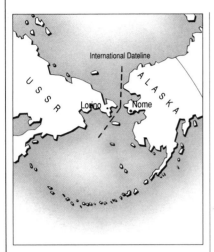

Crossing the Line *Leaving from Nome in Alaska, the* Rainbow Warrior *sails west across the international dateline to the USSR where she lands crew members at Lorino, the site of a whaling station and mink farm.*

Surprise Landing *Through the mist, Greenpeace inflatables approach the shore at Lorino, to the astonishment of the staff at the whaling station.*

Five people – Christopher Cook, David Rinehart, Ron Precious, Nancy Foote and Barbara Higgins – had elected to go ashore in three inflatables to film the whaling operation. David Rinehart recalls: "The five of us went ashore to film and pass out literature to incredulous Inuit workers who were unloading coal from an old barge. It was a beautiful morning; as the fog lifted we were able to photograph more and more of the whaling station and the animal pens above it."

Suddenly there was the sound of engines, and a small caravan of military trucks came rumbling down a slope. A squad of soldiers spilled out. The realization of their situation hit Nancy Foote hard: "I really was standing on Soviet soil without permission, surrounded by soldiers, facing possible imprisonment or a firing squad."

All five were arrested, as was Pat Herron when he came from the *Warrior* in an inflatable to deliver a radio so that Greenpeace's Russian translator, an East German defector named Wolfgang Fischer, could communicate with the Soviet commander. After an interrogation and a strip search, the protesters were held in the nearby village of Loren before being taken by truck and helicopter to a military barracks.

In the meantime the *Warrior*, which had filmed the incident, made a dash for the open sea carrying the vital film. She was already being buzzed by a military helicopter, and her radar screen showed that there were two ships approaching fast. Near the 19-kilometre (12-mile) territorial border, a Soviet gunboat closed in, accompanied by helicopters that circled overhead trying to get the boat to stop.

Willcox steered the *Warrior* expertly, refusing to give up. But when it appeared that they would soon be boarded, crew member Jim Henry volunteered to take the film of the whaling centre back to Nome in one of the fast inflatables. Soon helicopters and high-speed boats were in pursuit and the chase continued for the next two hours.

Arrested *Viewed from a Greenpeace inflatable waiting offshore, the scene (above) shows the Soviet military arriving to seize the protesters.*

Too Close for Comfort *A Soviet vessel gains on the* Rainbow Warrior *in an attempt to stop the Greenpeace protesters before they leave Soviet waters (left).*

August 12, 1982
Forty US Coast Guard vessels, using high-powered water cannons, broke up a blockade of small boats seeking to prevent the docking of the first US Navy Trident submarine at its base in Bangor, Washington State. Fifty-three people were arrested.

August 31, 1982
In protest at the failure of the US Environmental Protection Agency to act in the best interests of the environment, two Greenpeace activists – John Willis and Jamie Graeger – descended sheer cliffs to within 15 metres (50 feet) of the thundering waters of the Niagara Falls to hang a large banner reading "EPA – Protecting the Environment or Industry?". They were later charged with disorderly conduct.

Eventually a helicopter swooped so close that the waves whipped up by its rotors almost turned the boat over and Henry was thrown into the cold sea, narrowly avoiding the propeller as the inflatable spun wildly. As the crew of the *Warrior* watched through binoculars, they saw him being winched onto the Soviet helicopter, which then took him ashore to join the other detainees. The *Warrior* charged after the inflatable. Bruce Abraham, the third mate, leaped into the craft so it could be pulled aboard, breaking his ankle in the process. To their delight, they found the valuable campaign film hidden in the bottom of the boat.

Now the *Warrior* found herself being chased across the Bering Sea, pursued by a gunboat and other vessels. Helicopters fired flares across her bows and, on three occasions, the gunboat came close to the *Warrior*'s stern and ordered her to stop. A Soviet freighter cut across her bow in an effort to fence her in but Peter Willcox outmanoeuvred her. The chase lasted for 10 hours before the *Warrior* was finally free.

DEADLY INVISIBLE NETS

Before heading out on her Siberian mission, the *Rainbow Warrior* left Vancouver on June 12, 1983, to confront the 172-boat factory fishing operation of the Japanese gillnet fleet. These boats set huge 15-kilometre (9-mile) monofilament nylon nets to catch salmon. As the nylon is invisible to wildlife, some 8,000 Dall's porpoises and 250,000 seabirds, along with unknown numbers of sealions and seals, die in the nets each year. The fishing takes place in US waters, under permit from the government.

Greenpeace's action was designed to draw attention to the situation and to call for the nets to be modified. The campaigners believe that weaving air-filled tubing into the nets would make them more visible and thus reduce the dolphin kill by half.

The first encounter was with the *Yahiko Maru* 320 kilometres (200 miles) northwest of Attu, the westernmost island of the Aleutian chain. In the following two weeks the *Warrior* shadowed the fishing fleet and was herself shadowed by a Japanese patrol boat and a US reconnaissance plane. Greenpeace divers, who took to the water to search the nets, found only one dead porpoise but were able to disentangle many seabirds and release them alive.

Gales at the beginning of July forced the *Rainbow Warrior* to seek shelter in the lee of Attu island. When she relocated the *Yahiko Maru*, the crew launched their inflatables and positioned them under the stern, preventing the boat deploying its gillnets and causing it to lose a night's fishing. In a shouted exchange, the captain of the gillnetter stated that he was impressed with the protesters' stamina and persistence and that he understood and respected the Greenpeace position. The lead Zodiac succeeded in placing an "I support Greenpeace" sticker on the stern of the gillnetter.

***Gillnet Casualty** Greenpeace campaigner Barbara Higgins (right) holds a dead seabird, one of the hundreds of thousands that become entangled in gillnets every year.*

***Hounding the Culprits** Inflatables from the* Rainbow Warrior *follow hot on the heels of a Japanese fishing vessel to try to prevent it deploying the gillnets.*

October 2, 1982
Greenpeace Canada's president, Patrick Moore, testified against the Canadian seal hunt before the Court of International Justice for Animal Rights, the judicial arm of United Animal Nations, an organization dedicated to the cause of animal rights worldwide.

December 2, 1982
Sirius arrived in Swedish waters as part of a protest aimed at preventing the nuclear waste ship Sigyn taking its cargo to France for reprocessing.

After several tense days, the Greenpeace people being held on shore were at last prepared for release. They were not to be imprisoned, thanks to widespread press coverage and telephone callers who had blocked the lines of Soviet embassies around the world. Instead the protesters were to be handed over, on board the *Warrior*, to the mayor of Nome, who had been authorized by the US State Department to accept responsibility for them. The Greenpeace ship rendezvoused with five Soviet ships, including two warships, off St Lawrence Island in the Bering Sea, just east of the international dateline, and the handover was completed by the light of the midnight sun.

The mission was front-page news in London and New York – the most publicized Greenpeace event to date. In England, Executive Director John Frizell had a debate with a Soviet official on television, and in Canada the incident created a ream of news clippings.

OPPOSING FORCES

FOR THE SIRIUS, *1983 had begun with a bang. On January 5, while manoeuvring into France's Cherbourg harbour to block the arrival of the* Pacific Crane, *which had spent nuclear fuel rods from Japan on board, she was bombarded with tear gas grenades and boarded by military police, while more than 200 armed riot police lined the quayside. The ship's anchor chain was cut and equipment was smashed. "The deck was holed and everyone was blinded with tear gas," crew member Pierre Gleizes told reporters. "We tried to shout at the riot squads to stop, but they ignored our loud-hailers."*

Under Attack Sirius *crew members are arrested by French riot police.*

Give Her Back *Chained to the gates of the merchant navy building in Cherbourg, Greenpeace protesters demand the release of the* Sirius.

January 15, 1983
Marie Chapman and Knorrie McFarlane delayed the departure from Lyttelton, New Zealand, of the Hakurei Maru, *a Tokyo-registered vessel bound for the Ross Sea to search for oil. The two Greenpeace members attached themselves to the ship's mooring line, using modified mountaineering equipment, hanging a banner saying "Antarctica At Risk, No Exploitation".*

January 25, 1983
Norway announced it would end the hunting of seal pups, following the threat of an EEC ban and a global campaign by Greenpeace and other environmental organizations. The hunting of adult seals would continue, but the number of ships used would be reduced.

February 28, 1983
Under the banner "Stop the Murderers", a Greenpeace protester staged a dramatic lobbying action outside a meeting of the EEC Council of Environmental Ministers (below). The council later voted for a voluntary, rather than legal, ban on the trade in skins or products of harp seal pups below the age of three months. This ban was instituted on October 1, 1983, and ran for two years.

The ship was detained for about a week, when it was released without penalty. Public pressure certainly played a part in this. About 3,000 demonstrators turned up at one protest in Cherbourg, and there were supporting actions around the world. In Sydney, Australia, three members of Greenpeace chained themselves to the doors of the French consulate; in Copenhagen seven members were arrested after blockading the entrance to the French Embassy.

The following month, Greenpeace's years of campaigning to prevent the dumping of low-level radioactive waste at sea appeared to have paid off. The 56-member London Dumping Convention, the only dumping agreement that takes in all the oceans, voted by an overwhelming majority for a two-year moratorium on such disposal, pending a technical review to be carried out by an international group of scientists. (The moratorium was later extended indefinitely.)

This resolution, which Britain voted against, was not legally binding, however, and the UKAEA announced they were planning a dump some 50 per cent larger than that of the year before. In addition, they had adapted the waste carrier *Atlantic Fisher* to allow it to release its cargo below the waterline, making it "Greenpeace-proof". In response to this news, on June 17, four major transport unions in the UK agreed to black the transport of all radioactive waste by rail, road or sea.

The pressure of protest intensified as the date for the dump approached. On July 3, the Greenpeace boat *Cedarlea* travelled to Sharpness, near Bristol, where the waste was to be loaded. The ship supported a demonstration in the town and awaited developments there. Twenty-two protesters from Spanish and Belgian environmental groups were arrested on July 6 after 10 of their number chained themselves to the UKAEA headquarters in London.

On July 8, the day on which it should have begun its dump, the *Atlantic Fisher* was still in harbour at Barrow without a crew or its cargo. On July 11, there were demonstrations at British embassies and consulates in many European countries; 150,000 people in Spain marched against the dump, and telegrams of support for Greenpeace arrived from scientists, environmentalists and trade unionists around the world. As a result this dump was postponed indefinitely. But Britain still reserves the right to dump radioactive waste at sea in the future.

Another nuclear campaign, which proved to have serious implications for the organization, was that against the discharge of radiation from the nuclear reprocessing plant at Sellafield into the Irish Sea.

The scene of the world's first major nuclear accident, in 1957, Sellafield had been the subject of intense controversy ever since, particularly because of its appalling accident record and the extent of the pollution it was causing. Its outflow pipe was discharging more than 10 million litres (2.2 million gallons) of radioactive water into the sea every day. Greenpeace was determined to stop it.

During a tour by the *Cedarlea* in June, around areas of the west coast of the British Isles most affected by this pollution, Greenpeace divers confirmed the exact position and dimensions of the outflow pipe. They could now set to work on a device that would plug it.

March 17, 1983
Sirius in Norway blockaded a sealing ship and prevented it leaving Ålesund harbour. The Norwegian sealers were setting off to take part in the annual slaughter of 127,000 seals in the Jan Mayen and White Sea areas of the Arctic Ocean. The crew of the Sirius received a message of support from the Norwegian ethnologist and explorer Thor Heyerdahl.

LOST WASTE FROM SEVESO

On July 10, 1976, a cloud of toxic chemicals settled over an entire district of Seveso in northern Italy, after a factory explosion. Seven years later, 41 barrels of dioxin-laced detritus from this accident went missing.

Hoffman La Roche, owners of the Seveso plant, sent the barrels to the German company Mannesman for disposal, but they disappeared en route. Greenpeace discovered this and was responsible for prompting a search throughout Europe.

Eight months later, an investigative journalist tracked the barrels down in a disused French slaughterhouse at D'Anguilcourt-Le-Sart, near St Quentin. The sub-contractor responsible, Bernard Paringaux, received an 18-month suspended sentence and a £20,000 fine. The waste was finally burned in a high-temperature incinerator in Basel, Switzerland, in 1985.

On April 14, 1983, Greenpeace staged a piece of effective "street theatre" by making mock-ups of the 41 missing barrels and returning them to the administrative headquarters of Hoffman La Roche at Basel. The campaigners blockaded three of the building's entrances before being removed by police.

Lethal Waste Protesters stage a death scene with mock-ups of the Hoffman La Roche barrels.

April 6, 1983
Off the coast of California, three Zodiacs, launched from the Rainbow Warrior, *interrupted the schedule of the seismic survey vessel* Western Glacier, *in protest at the development of offshore oil reserves. The ship's survey permit was subsequently suspended, as it turned out to be operating in waters not authorized for exploration.*

Underwater Search *A Greenpeace frogman* (right) *dives in the Irish Sea off Sellafield to locate the radioactive outflow pipe.*

Monitoring *Prevented from plugging the pipe, Greenpeace continues to take samples of the waste water being discharged from the nuclear reprocessing plant.*

May 10, 1983
The Swedish freight ship Sigyn *had to force her way through a Greenpeace blockade of nine small boats at the entrance to the port of Barsebaeck in Sweden. The* Sigyn *was there to load two containers of spent nuclear fuel, destined for Cap de la Hague.*

The boat returned to the area in November and, on the 14th, four divers taking samples of silt near the pipe were contaminated by an oily radioactive slick that poured from it. The dose of radiation the men received was 50 times the normal background level. Over the next four days, the prevailing winds blew the slick, containing highly radioactive debris, onto the beach, and 40 kilometres (25 miles) of the coast had to be closed to the public for eight months.

When divers returned on the 18th to plug the pipe, they discovered, after a 17-hour search, that British Nuclear Fuels Ltd (BNFL) had welded two metal rods to the diffuser at the end of the pipe, making it impossible for Greenpeace to use its custom-built bungs.

As a result, the action was called off, but this did not prevent BNFL pursuing Greenpeace through the courts. BNFL had previously obtained an injunction in the High Court to prevent Greenpeace interfering with the pipeline, and in response to this latest action they applied

Test-Ban Treckers Brian Fitzgerald (far left), Harald Zindler, Jon Hinck and Ron Taylor cross the Nevada test site.

for the sequestration of all Greenpeace assets, in Britain and overseas. Greenpeace, in turn, filed notice to contest the BNFL claim and planned to file a counter claim for the contamination of its divers.

On December 1, after three refusals to give an assurance that it would not interfere with the pipeline, Greenpeace was found to be seriously in contempt of court and fined £50,000 plus costs. The sequestration application was set aside.

On January 13, 1984, BNFL were granted a permanent injunction against Greenpeace stating that the protest group was not allowed to interfere with the pipeline but could continue taking samples from the area. The High Court also accepted the £36,000 raised by public donations as complete payment for the earlier fine.

The matter did not stop there. The oily radioactive discharge that the Greenpeace divers had encountered was caused by a combination of operator error and equipment malfunction, resulting in an emission well above the legal level set for the plant. The event might have gone unnoticed if the divers had not been there at the time.

There were obvious legal implications. The matter was referred to the Director of Public Prosecutions and, at a hearing in July 1985, BNFL were found guilty on four criminal charges and fined a total of £10,000 – the first public utility in the UK to be charged in this way.

ACTION FOR PEACE

Nineteen-eighty-three also saw Greenpeace actively campaigning against the testing of nuclear weapons, calling for a resumption of negotiations between the US, Britain and the USSR aimed at achieving a Comprehensive Test Ban Treaty (CTBT), which would end nuclear tests by these nations for all time.

The first target was the Nevada test site, where the USA and the UK had conducted more than 600 nuclear blasts. Four Greenpeace activists – Harald Zindler, Ron Taylor of England and Americans Jon Hinck and Brian Fitzgerald – invaded the site in April 1983.

"We went in on the north side of the test site, which is nearly the size of Rhode Island," says Fitzgerald. "We were working off mining maps and some topographical maps and satellite maps. We tried to avoid

May 25, 1983
The Greenpeace ketch Aleyka (above) began two years of active campaigning on the east coast of North America by blocking the discharge pipes of SCM Glidden Corporation in Baltimore harbour in Maryland. This was the first of a series of actions conducted against the direct discharge of toxic waste into rivers and coastal waters.

June 4, 1983
The Aleyka joined a flotilla of 35 boats as part of a protest against At-Sea Incineration Inc.'s proposal to build a $100 million terminal at Port Newark, New Jersey. The terminal would store hazardous waste for subsequent incineration at sea.

August 20, 1983
Norway's biggest fishing company, Frionor, said it would stop whaling by 1986. It admitted it had lost orders worth more than £6 million since Greenpeace had organized a consumer boycott of Norwegian and Japanese fish exports to the US three weeks earlier.

areas that were marked as 'radioactive', and military high security…" Leaving their four-wheel-drive vehicle, they set out on foot, equipped with survival gear. "We carried Geiger counters and didn't detect anything very high. When we got to the target above Yucca Flats we took pictures of the craters and stayed hidden in the underbrush. We stayed four or five days, hearing the helicopters overhead at night. We waited for the test; it didn't come, and we were running low on water so we decided to pull it in. We walked down to the test site – it was like a moonscape – and went to where they were drilling for a test…We spent the night in jail. We think we delayed the test for a day."

TAKING FLIGHT

An equally bold protest took place in Germany. On August 28 the hot-air balloon *Trinity* flew over the Berlin Wall into East Germany, carrying pleas for peace and disarmament.

"The major powers, the USA, the USSR, France and the United Kingdom, jointly control the air space over Berlin," said a statement issued by Greenpeace, "and the city therefore provides a unique focus for demonstrating against these testing nations simultaneously."

Aboard the balloon were Gerd Leipold from Germany and John Sprange from Britain. "It took long preparation," says Leipold. "Two months of training here in Germany before we were able to fly. Then in Berlin we had to make sure no one detected us… We had to wait a week for the right wind. It was quite a nervous time – getting up at three each morning to check the weather."

When the right wind came, however, the 40-minute early morning flight went smoothly. "We got up in the air and all proper authorities, east and west, were informed. We asked that they use no violence. We crossed the Wall and rather quickly landed. About 10 or 15 minutes later … we got arrested and were treated rather well. They waited for orders and after about five hours we were expelled."

Risky Undertaking *The* Trinity, *Greenpeace's hot-air balloon (below), flies east across the heavily-guarded Berlin Wall.*

"Time to Stop Nuclear Testing" *Greenpeace climbers* (right) *scale the scaffolding around London's Big Ben clock-tower to hang their anti-nuclear message.*

August 27 – September 4, 1983 *Members of Greenpeace Australia joined 800 people for a blockade of the largest uranium mine in the world at Roxby Downs in the centre of the South Australian desert. There were 260 arrests including 50 Greenpeace people.*

The balloon, however, remained in custody. (On November 8, 1984, the activists were fined a total of DM 700 by a West Berlin court after being found guilty of importing "warlike material" into Berlin and illegally using radio equipment during the flight.)

Further imaginative protests on the same theme were to come the following summer. Greenpeace bought a bus from a film company for £3,000 and converted it to look like a London Transport double-decker, fitting it with a sliding roof and an extension ladder. On June 11, 1984, in the heart of London, two climbers – British teacher Ron Taylor and Renato Ruf, a mountaineer from Zurich – surprised the authorities by using this unique protest vehicle to reach the scaffolding surrounding the clock-tower of the Houses of Parliament, known as Big Ben, and clambering 55 metres (180 feet) up it. They remained aloft for 11 hours, hanging 2-metre- (7-foot-) high red letters across the face of the famous clock: "Time To Stop Nuclear Testing".

Two months later, on August 6, four Greenpeace climbers accomplished a similar feat in New York harbour, scaling scaffolding that encircled the Statue of Liberty. With the Greenpeace sailboat *Aleyka* standing by, the protesters– Sebia Hawkins, David and Mike Rapaport and Steve Loper – attached safety harnesses to the scaffolding and stayed for five hours. This time the banner, hung on Hiroshima Day, said: "Give Me Liberty From Nuclear Weapons. Stop Testing".

November 9, 1983
The Greenpeace yacht Vega *forced the USS* Phoenix (above), *a 6,000-tonne nuclear-powered submarine with nuclear weapons aboard, to halt her entry into Auckland harbour, New Zealand. The submarine was met by a fleet of 100 sailing boats, canoes and surfboards, and was forced to a standstill when the Greenpeace yacht attempted to cross her bow. The* Vega *was surrounded by a large contingent of navy and police boats. The skipper, Chris Robinson, and Ann-Marie Horne, veteran of the 1973 Moruroa voyage, were both arrested.*

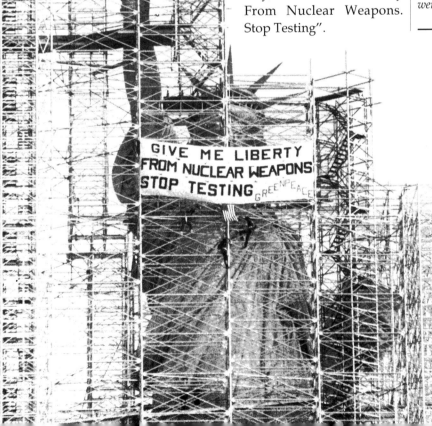

"Give Me Liberty from Nuclear Weapons. Stop Testing" Protest climbers at the top of the Statue of Liberty in New York harbour call for a halt to the US nuclear weapons test programme.

"She Too Tried to Save the
Whales" Copenhagen's Little
Mermaid becomes the latest recruit
to the whale campaign.

While extending the range of its campaigns, Greenpeace never lost its commitment to the whales and, in 1984, continued to draw attention to the whaling nations, principally Japan and the Soviet Union.

In many cities throughout the world, including Stockholm, Amsterdam and Bonn, demonstrators paraded outside Japanese embassies, but the action that gained the biggest media coverage was in Denmark. Greenpeace activist Michael Nielsen blindfolded the Little Mermaid statue in Copenhagen harbour with an American flag (in protest at US refusals to enact sanctions against Japan) and fashioned a harpoon out of cardboard and fitted it so that it appeared to pierce the mermaid's chest. He draped a bloody Japanese flag by her side and hung a sign on her saying: "She Too Tried To Save The Whales".

These symbolic protests were complemented by more direct actions when, in August, the *Sirius* was sent to intercept Soviet whalers on their way to the southern hemisphere.

"What we did was sail *Sirius* into the Mediterranean, and send two people to Turkey to stand on a bridge to spot the factory ship, and then we waited for the call that the whalers were on the way," says Nielsen. "One morning we saw a catcher boat, and we occupied it in the middle of the Mediterranean – climbing aboard in the early morning from two inflatables. One Greenpeace man, Spaniard Xavier Pastor, climbed up to the crow's nest. Maggie McCaw and Leo Snellink latched themselves to the rails. The Soviets had to stop the ship and we gave them messages. We tried to follow the factory ship out of the Gibraltar Strait but a military tugboat kept interfering."

FOCUS ON POLLUTION

But, throughout 1983 and 1984, the greatest number of actions was against toxic pollution. The principal targets of these campaigns were chemical plants discharging titanium dioxide waste into rivers and oceans. The actions began in March 1983, when divers from the *Cedarlea* attached a tube to the end of the wastepipe used by a company called Tioxide to discharge 800,000 tonnes of waste per year into the Humber estuary in England. The tube caused the waste to spray into the air in a giant fountain, drawing attention to this pollution hazard.

A string of actions was to follow against three German factories, owned by Kronos-Titan and Pigment Chemie, which between them were dumping more than a million tonnes of titanium dioxide waste, laden with sulphuric acid and heavy metals, in the North Sea every year. In August 1983, four Greenpeace divers and a flotilla of four Zodiacs operating from the *Sirius* forced the dump ship *Titan* to return to port fully laden when they formed a human chain in front of the vessel's bow. The protest was so effective that the West German government invited Greenpeace representatives to discuss the issue as a result.

At around the same time, actions in Norway delayed the laying of a pipeline by Titania, a subsidiary of Kronos-Titan, which was to discharge titanium dioxide waste into an area of natural beauty.

In September, four crew members from the *Cedarlea* chained themselves to the discharge pipes of the Belgian dump ship *Falco*, loaded with chemical waste from the titanium dioxide factory of NL Industries in Ghent, who were dumping some 300,000 tonnes of waste per year in the waters off the Dutch coast.

Four days later, 12 Greenpeace volunteers from the *Aleyka* plugged the discharge pipes from the Gulf and Western plant that were emitting 40 million litres (9 million gallons) of titanium dioxide waste a day into the Delaware River at Gloucester, New Jersey.

MOUNTING PRESSURE

Back in Germany, three fishing boats and two Greenpeace boats blockaded the loading dock of Kronos-Titan's factory at Nordenham, preventing two waste ships being loaded. A larger blockade followed in February 1984, when 54 fishing boats joined the *Sirius* in a concerted action against the plant, backed by many environmental groups and the 50,000 members of the German National Fishermen's Association.

Pigment Chemie's factory in Duisburg was blockaded in May by a Greenpeace inflatable. Greenpeace climbers shinned up the loading pier and, 20 metres (66 feet) above the water, blocked it with a steel cage that was then occupied by Greenpeace protesters.

In Canada on July 4, ten Greenpeace people were arrested after attempting to plug the discharge pipe of Tioxide of Canada in Tracy, Quebec, which was pouring between 200 and 600 tonnes of concentrated acid each day into the St Lawrence River. Charged with criminal mischief, the ten were fined $250 each and placed on probation.

The year was to end with the announcement that Kronos-Titan and Pigment Chemie would stop dumping in the North Sea in 1988, and that a recycling plant for both companies would be built.

January 1984
At a meeting of the Antarctic Treaty Consultative Parties in Washington DC, Greenpeace presented evidence of the damage being caused by the building of the French airstrip at Pointe Géologie Archipelago off the Antarctic coast. This forced the French to re-examine the Environmental Impact Assessment that had been carried out, but the committee's conclusion that the project should be abandoned was kept secret.

On July 29, *six Greenpeace members dressed as emperor penguins scaled the offices of the Terres Australes et Antarctique Françaises (TAAF) in Paris, demanding the release of the committee's report. Another group of penguin demonstrators* (above) *later that year delivered 20,000 postcards sent by members of the public in protest at the airstrip's construction.*

Toxics Action *Steve McAllister and Jim Henry block the waste discharge pipes of the Monsanto chemical plant in Boston, Massachusetts.*

February 6, 1984
Police from the Serious Crimes Branch of New Scotland Yard arrived with a search warrant and ransacked the London offices of Greenpeace and Friends of the Earth. The search was prompted by the two groups' disclosure of a leaked government document relating to the disposal of plutonium-contaminated waste in the Atlantic.

February 29, 1984
The Maritime Tribunal of Rouen fined Jon Castle 5,000 francs for breaking an injunction at Cherbourg in 1980. Willem Beekman was fined 6,000 francs and given a suspended six-day prison sentence for not leaving territorial waters when ordered to do so in June 1983, during the gypsum sludge campaign off Le Havre.

March, 1984
The 1984 Canadian seal hunt was severely restricted by bad weather and was on a smaller scale because the EEC ban on imported seal goods had removed the major consumer market. Magdalen Islanders destroyed a helicopter belonging to the International Fund for Animal Welfare, after its crew requested an emergency refuelling there. A Greenpeace aeroplane was seized by the authorities and pilot Bruce Jaildagian and cameraman Ron Precious were arrested for violation of the Seal Protection regulations.

March 6, 1984
Greenpeace demonstrated outside the missile range in Cold Lake, Canada, where the first test of a US nuclear delivery system in Canadian airspace, involving four unarmed cruise missiles, was successfully completed.

March 8, 1984
Four Greenpeace members were arrested for trespass after "occupying" a toxic waste pond at a Chevron USA petrochemical plant in Richmond, California, and closing three valves that release the waste into a nearby creek. The action was part of a two-day nationwide Greenpeace campaign aimed at drawing public attention to "the toxic time-bomb".

THE MONT LOUIS AFFAIR

On August 25, 1984, the 5,800-tonne French freighter *Mont Louis* sank 9 kilometres (12 miles) off Ostende, Belgium, after colliding with a Dutch cross-channel ferry carrying 935 passengers. The possible consequences of this accident might have been overlooked had not Greenpeace discovered that the freighter's destination was Riga in the Soviet Union and that it was a sister ship to the *Borodine*.

For 10 years the *Borodine* had been shipping uranium hexafluoride – the basic raw material from which nuclear fuel is made – to the USSR where it was enriched and then shipped back for use in French nuclear reactors.

On Sunday August 26, Greenpeace Paris released what information it had about the *Mont Louis* to the media and suggested that the ship's cargo was uranium. Later that day this was confirmed by the French Democratic Federation of Labour (CFDT).

The full details of the cargo were not known for 24 hours. It was then revealed that the ship was carrying 200 tonnes of uranium hexafluoride in 30 reinforced containers. The barrels belonged to the French Compagnie Générale de Matières Nucléaires (Cogéma) and the Belgian power company Synatom. If punctured they would produce a violent chemical reaction.

It was also revealed that this was the first nuclear shipment to be carried on the *Mont Louis* and that the crew were not experienced in handling such a hazardous cargo.

The Greenpeace vessel *Sirius* arrived in Ostende on September 10 and a press conference was held on board featuring Jim Slater, general secretary of the UK National Union of Seamen, who called upon the International Maritime Organization to ban such shipments until stricter international safety regulations had been introduced.

Two days later the *Sirius* sailed close to the wreck of the *Mont Louis* and observed empty barrels floating in the sea; a minesweeper and a tugboat were spraying detergent on an oil slick that had formed around the boat, and six warships were patrolling the area.

Vital Warning
On October 4, the last barrel was extracted from the wreck. Fortunately, none of the containers was damaged, but the whole incident drew attention to the international trade in uranium and plutonium and provided a warning of the potential dangers. As Roger Bradley, a Lloyd's underwriter who pioneered nuclear cargo insurance, commented at the time: "We are damned lucky she wasn't hit by an oil tanker. The physical impact would have burned up the *Mont Louis* before she sank and the uranium cargo would have exploded."

Radioactive? An empty barrel from the Mont Louis *is washed up on the beach at De Haan, near Ostende.*

"The Sea is Not a Dustbin"
French Greenpeace members protest
against the chemical companies' use
of the North Sea as a dump site.

July 16, 1984
Six days of protests against uranium
mining ended when the cargo vessel
Clydebank, carrying 32 containers of
"yellow cake" from the Ranger Mine,
finally left from Darwin harbour,
Australia. (Yellow cake, a mixture of
uranium oxides that contains 85 per
cent uranium by weight, is the raw
material for all the succeeding processes
in the nuclear fuel cycle). Loading of the
ship had been delayed by a crowd of 60
Greenpeace members plus local
residents, of whom five were arrested.
Two other Greenpeace members were
later arrested after they had stowed
away aboard the Clydebank, which was
accompanied out to sea by a spectacular
flotilla of local boats and the Vega,
covered in anti-uranium banners.

"Recycle the Yellow Mud" A
fishing boat (left) bears a banner
calling for an alternative to the
dumping of gypsum sludge in the
Seine estuary.

"S O S. Let's Save the Sea" A
French Greenpeace sticker (below)
calls for support in protecting the
oceans from pollution.

Another focus of Greenpeace's concern was the dumping every year of 2 million tonnes of gypsum sludge – a residue from the production of phosphate fertilizers containing poisonous cadmium – into the Seine estuary by three companies: Cofaz, and the government-owned Rhône-Poulenc and APC. The bay had once been a rich fishing ground, but now the sludge was promoting the spread of a phytoplankton that is toxic to marine life.

The opening salvo of this campaign came in June 1983, when the *Sirius* and a flotilla of 40 fishing boats harassed the dump barges as they approached the dump site 11 kilometres (7 miles) off Le Havre. One of the 20,000-tonne barges was sprayed with paint, occupied and immobilized by Greenpeace activists; as a result, the *Sirius* was arrested by the French Navy and escorted out of French territorial waters.

Giddy Heights The dizzying view from the chimney top (above) *as the second climber nears his goal.*

Greenpeace was equally concerned about the effects of atmospheric pollution and, in particular, the growing threat posed by acid rain.

In a spectacular protest action, on the morning of April 2, 1984, Greenpeace teams simultaneously climbed smokestacks in Belgium, West Germany, Austria, the United Kingdom, Denmark, the Netherlands, France – and Czechoslovakia. From each one they hung a banner with a single letter on it, so that a composite photograph of all the smokestacks would show the banners reading "STOP", twice over.

Preparations for the protest in Czechoslovakia were made in West Germany. When the time came, the Greenpeace activists – including Lena Hagelin of Sweden – drove into Czechoslovakia and secretly painted their banner in a wood.

"We went into the factory and went right up the chimney and the workers just stood staring at us," said Hagelin. "The secret police came out and thought we were terrorists. And so they started to shoot - [a bullet] hit just a hand's length from my head on the chimney! I don't

know if they were real bullets, but they started to shoot and we ran up quick as we could! But we wanted to hang that banner, and we didn't get scared till afterwards… They stopped shooting and we unfolded the banner. Then the fire brigade came. They took hoses and began climbing up. We didn't want to mess with them, and we already had a photograph of us taken for the protest picture, so we went down. There were police cars and we were taken to the police station. Greenpeace put pressure on the various embassies, and the embassies put pressure on Prague, and so we were expelled the same day with a small fine."

Although no one was hurt, the shooting incident did not bode well. Greenpeace remained dedicated to the principles of non-violent direct action but was often subject to rough treatment at the hands of the

"S T O P" Daring Greenpeace climbers scale industrial smokestacks across Europe and hang banners that together spell out their demand for an end to the atmospheric pollution that is causing acid rain.

Made It The Austrian climber, armed with a banner, reaches the top of his chosen smokestack.

July 29, 1984
Four Greenpeace inflatables and 15 canoes surrounded a vessel bringing 70 tonnes of Norwegian shrimp into Portland, Oregon, to protest at Norway's intransigent whaling policy. Norway exports $70 million worth of seafood products to the US every year.

October 25, 1984
Three Greenpeace "penguin protesters" ended their occupation of the Norwegian vessel Polarbjørn at Le Havre after 56 hours. The ship, under charter to the French Antarctic Authorities, was to transport provisions and construction materials for the French airstrip in Antarctica. On December 5, the Vega attempted to prevent the ship from docking when it arrived to load cargo in Hobart, Tasmania, and it was again occupied by Greenpeace members.

(As a result of international pressure, work on the airstrip was substantially delayed until November 1987, when the French government announced that construction would resume immediately, would last for five years and would cost 100 million francs.)

TURTLES IN TROUBLE

The eight species of sea turtle are among the most endangered marine animals in the world.

Greenpeace's sea turtle campaign, which began as a little beach patrol in Florida, became a multi-faceted international campaign in 1984, operating in at least 10 countries in the Caribbean and investigating illegal trade in turtle products in the South Pacific.

The turtle campaign has several main strands:

• Tens of thousands of turtles a year die by being entangled in shrimp fishing nets. Greenpeace lobbied for the installation of simple trapdoors in such nets in the US. This became mandatory in 1988 and Greenpeace is now using this legal precedent to call for the use of these in shrimp

Turtle Care *Wolfgang Fischer kneels next to a giant leatherback turtle.*

fishing operations in Mexico, Australia and other parts of the world.

• Greenpeace is establishing beach patrols and building hatcheries in order to protect clutches of eggs at nesting sites in the US, Greece, Mexico, Costa Rica, French Guiana and many other countries.

• The international illegal trade in tortoise-shell and other sea turtle products is being investigated. Cuba is the largest exporter of tortoise-shell in the world (it is not a signatory to the Convention on the Illegal Trade in Endangered Species [CITES]) and Japan is the largest importer of these and other endangered species.

• The campaign opposes plans for the establishment of sea turtle ranches and farms, which take eggs and adults from the wild every year to restock. The existence of these commercial enterprises helps to promote the international trade in wild animals.

"Clean Air, Not Hot Air"
Greenpeace expresses frustration at delegates' inability to solve the problem of acid rain at a pollution conference in Munich.

November 2, 1984
Greenpeace UK won the first case ever brought under Section 6 of the Endangered Species Act. Reg Bloom, director of Knowlesley Wildlife Safari Park, pleaded guilty to four charges of taking dolphins and keeping them, against the provisions of the act.

Protest Jump *Robin Heid parachutes from the top of the smokestack at Ohio Power's generating plant in Cheshire, Ohio, one of the worst emitters of sulphur dioxide in the US.*

authorities. Nowhere was this more true than in France, where the government had become increasingly hostile to Greenpeace. The initial wound to France's pride inflicted by McTaggart's voyages to Moruroa in the early 1970s had been exacerbated by Greenpeace's continued campaign over the years. Now the French learned that Greenpeace was intending to send its flagship, the *Rainbow Warrior*, to Moruroa in an attempt to end nuclear testing there once and for all. It was to prove the final straw.

Greenpeace declared 1985 as "the year of the Pacific". It was also to be the year of disaster.

CHAPTER ELEVEN

THE END OF THE RAINBOW?

THE RAINBOW WARRIOR had never looked better. She now had majestic sails, fitted during another major renovation in Jacksonville, Florida. Freshly painted, the doves of peace shone whiter and brighter than before, the rainbows arching completely round her bow. She had been equipped with a new radio and radar. The bridge had been rebuilt, the engines overhauled, and Greenpeace's famous flagship could now clip along at 10 or 11 knots, using half as much fuel as before.

Resplendent *The* Rainbow Warrior *in full sail.*

The job had taken several months and cost well over $100,000, but it had all been worth it. The *Warrior's* lines responded brilliantly to the power of the wind. "We were amazed," says Dutch crewman Henk Haazen, "at how well she sailed."

Peter Willcox was at the helm, complemented by engineer Davy Edward, a Yorkshireman and former merchant seaman, and first mate Martin Gotje, a Dutchman who for years had sailed on the *Fri*. The crew also included representatives of Switzerland, Denmark, Germany, Ireland and New Zealand.

American Steve Sawyer, a member of the Greenpeace International board, directed the Pacific campaign. He had a complex task on his hands. In all it would be a 32,000-kilometre (20,000-mile) expedition, including stops in Hawaii, the Marshall Islands, Kiribati, Vanuatu and New Zealand, before going on to Moruroa itself in August, where the protest would begin in earnest.

The *Warrior* left Jacksonville on March 15, 1985, and sailed to Honolulu. There she picked up Sawyer and Fernando Pereira, a Dutch photographer whose job it was to document the voyage, and then headed for Rongelap in the Marshall Islands, where Greenpeace was to carry out what had come to be known as "Operation Exodus".

Between 1946 and 1958 Rongelap had been dusted by fallout from at least five US nuclear tests and, despite assurances from the Americans, the islanders were convinced that their home was still unfit for habitation. Aware that the *Warrior* was coming to the Marshall Islands, Senator Jeton Anjain, the Rongelap representative in the Marshallese

---

January 1985
Greenpeace, along with 21 other environmentalist and animal welfare groups, called for an international boycott of Japanese Air Lines in protest against continued Japanese whaling. Actions took place at JAL offices in Madrid, Stockholm, Hawaii, Paris, Vienna, Vancouver, Auckland, Chicago, San Francisco and Seattle. In one protest, a whale holding a placard saying "Say Goodbye To My World" parodied the JAL slogan (above).

This month, Canadian Greenpeace activists made three attempts to catch a cruise missile during test flights from the military base at Cold Lake, northeast of Edmonton, by suspending drift nets from weather balloons in the predicted flight path. The missiles, designed to fly below 60 metres (200 feet) to avoid radar detection, eluded the nets by flying at an altitude of 150 metres (500 feet). The aim of the action was to highlight demands for a cancellation of the four-year missile test programme.

January 11, 1985
Greenpeace representatives delivered a petition to the British government signed by 250 members of parliament from nine countries, calling for a reduction in emissions of sulphur dioxide, a major cause of acid rain.

Mercy Mission *In May 1985, the* Rainbow Warrior (below) *arrives at Rongelap Atoll. The islanders, their belongings packed* (top right), *are prepared to leave their home, contaminated by radioactive fallout, and face an uncertain future for the sake of their children* (below right).

BACKGROUND TO RONGELAP

Rongelap is a tiny Pacific island, about 145 kilometres (90 miles) east of Bikini. The Americans had staged 66 nuclear tests at Bikini and nearby Eniwetok between 1946 and 1958, and at least five of these had resulted in fallout upon Rongelap. Two decades later, the islanders were feeling the effects: high rates of thyroid cancer and leukaemia; children born with deformities or suffering retarded growth.

The biggest blast of all was that of an H-Bomb, code-named Bravo, which took place on March 1, 1954. At 15 megatons it was more than 1,000 times the strength of the bomb dropped on Hiroshima.

Under the Mushroom Cloud

On Rongelap the islanders were blinded by a flash that became a ball of fire, which gradually formed into a giant orange mushroom cloud. The sky turned the colour of blood. Seconds later, the sound of a deafening explosion hit them, followed immediately by a tornado-like wind that twisted trees and wrenched the roofs off their houses. Later, the island was dusted by a strange "snow" composed of pulverized coral and debris.

"We were very curious about this ash falling from the sky," Mayor John Anjain, told a reporter. "Some people put it in their mouths and tasted it. One man rubbed it in his eye to see if it would cure an old ailment. People walked in it, and children played in it." Soon they felt a terrible itching of the skin and severe nausea and many lost their hair. The drinking water turned yellow and bitter.

After three days, the Americans had evacuated them, but they had returned them to their homes three years later, with assurances that the island was safe. Now Anjain and his brother, a senator to the Marshallese Parliament, felt passionately that the people should leave for good. When the Americans ignored their request for help, Senator Jeton Anjain approached Greenpeace.

Operation Exodus Rongelap, where the health of many adults and children (right) has suffered as a result of fallout from US nuclear tests, lies in the Marshall Islands (below). On their way there, the crew of the Rainbow Warrior (bottom right) were joined by three Marshallese men, whose presence was to ease the arrival of Greenpeace at the Pacific island, and by Senator Jeton Anjain (bottom left), whose discussions with Steve Sawyer set in motion the evacuation of the contaminated atoll.

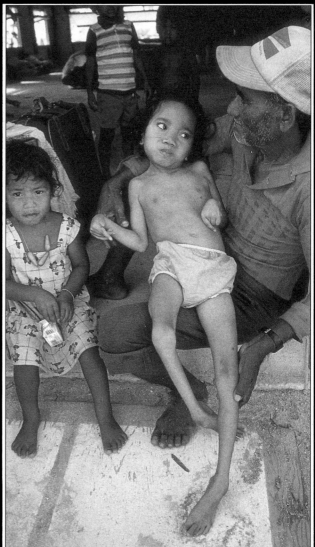

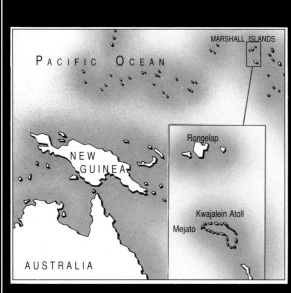

All Aboard *As loading begins,* *three children of Rongelap sit* *bemused on the deck of the* Rainbow Warrior *(left). The* *islanders, as well as some 100* *tonnes of belongings and building* *materials from dismantled homes,* *were ferried out to the Greenpeace* *ship in small boats (below) to* *make the 14-hour crossing to the* *uninhabited island of Mejato and a* *new life.*

Fernando Pereira The Dutch/ Portuguese photographer joined the Rainbow Warrior *almost by chance. While the ship was at Rongelap, the crew celebrated his thirty-fifth birthday.*

February 25, 1985
Greenpeace activists blockaded the Essi Flora *in the port of St Nazaire, France, for three days. The* Essi Flora *routinely transported thousands of tonnes of organic lead compounds used as an anti-knock additive in petrol. The action also highlighted the fact that France had been a major objector to an EEC initiative to introduce lead-free petrol.*

February 26, 1985
In protest at the nuclear waste discharges from Sellafield, Greenpeace members tipped five tonnes of radioactive mud, collected from the Ravenglass estuary, onto the steps of the Department of the Environment building in London.

March 5, 1985
Greenpeace climbers Joe Simpson and Paul Drury (below, seen here with Peter Wilkinson) scaled a 73-metre (240-foot) chimney at the Tioxide UK Ltd plant on the Humber estuary and, four days later, an international team of nine Greenpeace members blocked the plant's waste discharge pipeline, using a specially-made metal cap. The company had refused to accept an EEC directive calling for the reduction and eventual abolition of discharges of titanium dioxide waste.

(In August, Greenpeace was served with a High Court injunction to prevent any further actions.)

Nitijela (parliament), had approached Greenpeace with a radical proposal: that the boat evacuate the entire population to a safer island called Mejato, 195 kilometres (120 miles) away.

The *Rainbow Warrior* arrived at Rongelap on May 17, 1985 and, in the days that followed, the crew helped the islanders dismantle their homes, gather the furniture and ferry some 100 tonnes of supplies and materials out to the Greenpeace ship.

There were four trips to Mejato in all, over a period of 10 days, to relocate about 300 residents. By May 30 the operation was complete, and Fernando Pereira's pictures of the mission were soon being carried by the wire services all over the world.

WAITING FOR THE *WARRIOR*

In Auckland, Greenpeace members Elaine Shaw and Carol Stewart could hardly believe that the *Warrior* was at last on her way to New Zealand. They planned welcoming rallies and coordinated the press, who were showing a keen interest in the expedition.

It was no secret that the *Warrior's* chief destination on the Pacific voyage was Moruroa. Greenpeace had sent a telegram to French President François Mitterand telling him of the protest. It did not, however, mention an additional fact that Mitterand, acutely sensitive to independence movements in the Pacific territories, would have found especially upsetting: the possibility that Polynesians would join the protest, and would launch outrigger canoes from the *Warrior* in the waters around Moruroa.

But the French learned of the idea by surreptitious means. At around the time that the *Warrior* was in Honolulu, an agent of the French secret service had infiltrated the Greenpeace office in Auckland. She called herself "Frédérique Bonlieu" and she arrived on April 23, 1985, posing as an ecologist.

Frédérique—whose real name was Christine Cabon—was introduced to Greenpeace New Zealand by way of a letter from Jean-Marie Vidal, who had been involved in anti-nuclear activities in the Pacific. Jean-Marie said his friend Frédérique was coming out for a meeting on coral atolls and that she wanted to come down and meet Greenpeace people," says Elaine Shaw. "She was very standoffish. But she offered to do some work in the office and did a bit of translation."

Since the office operated on a low budget, relying mainly on volunteers, Frédérique's offer to help had been readily accepted. She folded newsletters, sealed envelopes, and sorted address labels. She also answered the telephone and read telexes. Messages had been flowing constantly from the *Warrior* since the Rongelap evacuation began.

Frédérique was also collecting maps of the area and photographs of many of New Zealand's beaches and harbours, checking hotel rates, and investigating where underwater equipment could be obtained and at what cost. All this material she sent to Paris, saying she had friends who were coming to New Zealand for a holiday.

What she was really doing, of course, was gathering information and laying the groundwork for French saboteurs who would soon arrive on an incredible mission: to sink the *Rainbow Warrior*.

AGENTS OF DESTRUCTION

The French sabotage was carried out by several teams of covert operatives. The first team arrived on New Zealand's northernmost coast by way of a yacht called the *Ouvéa*. On board were three French agents of the Direction Générale des Services Extérieurs (DGSE) – Gerald Andries, Jean-Michel Bartelo, and their leader, Roland Verge – along with Xavier Maniguet, a doctor specializing in treating victims of diving accidents. Their task was to smuggle explosives, diving gear and an inflatable Zodiac dinghy into New Zealand. Greenpeace had been employing the versatile Zodiacs for all kinds of non-violent protest ever since the French had used one to chase and board the *Vega* in 1973. Now, ironically, the French were to turn this manoeuvrable craft against Greenpeace in the most violent of ways.

French Spy *While working at the Greenpeace office, Christine Cabon was gathering intelligence.*

THE BELUGA

The *Beluga*, the new 24-metre (80-foot) Greenpeace riverboat, was baptized in Hamburg with clear water from the sources of the Rhine, the Elbe, the Thames, the Klar, the Seine and the Danube.

Built in 1961 in Bremen, West Germany, she was originally a fire-fighting ship. Purchased by Greenpeace in 1984, with the money from a huge fund-raising campaign, the ship was completely reconditioned by 40 volunteers who worked for 10,000 hours to transform her into an international action and laboratory ship for rivers, estuaries, harbours and coastal waters. She was

River Boat *The* Beluga *toured the waterways of Europe.*

renamed after the species of small white whale that was previously found in European rivers when the water was pure enough.

The aft part of the cabin was strengthened to enable the deck to carry a rigid-hull inflatable, and a hydraulic derrick was installed on the stern to lift the inflatable in and out of the water. The on-board laboratory was equipped with a computer and scientific equipment capable of identifying the chemical constituents of waste being discharged into water, and recognizing changes in water quality.

In the following three years, the *Beluga* monitored pollution in the waters of Europe, including the rivers Rhine, Elbe, Schelde, Seine, Weser, Mense, Thames, Humber, Severn, Tyne, Tees and Mersey, and off the coast of Denmark, Sweden, the UK and the Netherlands.

At Auckland airport, meanwhile, a second team arrived: Major Alain Mafart and Captain Dominique Prieur, posing as Alain and Sophie Turenge, a Swiss couple on their honeymoon. They rented a camper van, which they would later use to collect the equipment brought in by the *Ouvéa*. Next day, Colonel Louis-Pierre Dillais also arrived; he would oversee the whole operation.

As the various French agents were making their final preparations, the *Warrior* headed for meetings and rallies in New Zealand prior to setting out for Moruroa. In the afternoon of July 7, a cool Sunday, she was met by a welcoming flotilla of 30 boats as she made her way to Marsden Wharf in Auckland.

Hans Guyt, one of the directors of the Netherlands office, had joined the *Warrior* by now, and other senior Greenpeace members were flocking in from all over the world. Lawyer Peter Bahouth would soon arrive, along with Patrick Moore and Jim Bohlen, one of the founders of the Don't Make A Wave Committee who, now in his sixties, had recently rejoined Greenpeace in the Vancouver office.

The *Warrior* herself served as the centre of operations. On July 10, Sawyer's birthday, there was a celebratory dinner and, later, a meeting with the skippers of the "fleet" heading for the French atoll. Throngs of well-wishers milled about, including one man who introduced himself as François Verlon, a French "pacifist".

Verlon is now believed to have been another spy, whose real name is François Régis Verlet. He had arrived from Singapore the Monday before, bringing the total number of French agents involved to at least 10. Two others had arrived in Auckland on the same day as the *Warrior*.

THE FATAL ACT

Around 8.30 p.m. on July 10, Jean-Michel Bartelo tied up his inflatable at Marsden Wharf, put on his scuba gear and slipped into the water, making his way towards the *Rainbow Warrior*. He carried with him two packets of plastic-wrapped explosives. He tied one packet on top of the stern tube housing just forward of the propeller; the second one he fixed to the outer wall of the engine room. Bartelo set the timing devices, then swam back to his inflatable.

For some reason, he threw the outboard engine and his oxygen tanks into the sea before hauling the inflatable onto the shore and driving off in the camper van that Mafart and Prieur had parked nearby.

On board the *Warrior*, meanwhile, things had quietened down. After a long meeting, Sawyer together with Jim Bohlen and Patrick Moore headed for a hotel called The Surf Club outside Auckland, where all the trustees from Australia, New Zealand, Canada and the US were gathered for a regional meeting. Several crew members, including Guyt, Willcox, Gotje, Edward and Rien Achterberg, and photographer Fernando Pereira, remained on board having a drink in the mess.

The first bomb exploded at 11.38 p.m., sounding like a muffled thud, but powerful enough to lift those in the mess room off their seats. The ship lurched upwards and sideways. "It's the engine room!" Edward shouted. Dashing into the engine compartment he saw a hole the size of a car in the *Warrior*'s side through which water was pouring at the rate of six tonnes per second. As the *Warrior* keeled over, Willcox ordered everyone off the ship. The doctor on board, Andy Biedermann of Switzerland, went to check the cabins and pulled the cook, Margaret Mills, to safety. Gotje went below frantically searching for his girlfriend Hanne Sorensen; Pereira, too, rushed down below, perhaps to get his valuable cameras, which were on his bunk.

"There wasn't panic," recalls Hans Guyt, "just confusion. You could hear water pouring in. I went to the left side. Gotje was with me and Fernando was with me. I remember Fernando saying, 'She's sinking! She's sinking!' Water had already been coming on the deck. But still there was no panic. I went up one deck higher to help release a yacht moored alongside the *Rainbow Warrior* and get it away. I went down again and by now the water was knee-deep. I was standing on top of the stair leading to the lower accommodation, which was flooding, and I could see Gotje with a lamp groping for his girlfriend. I was hesitating as to whether to go down because my cabin was down there as well, and then the second explosion occurred, right under our feet."

The remaining crew jumped onto the wharf - all except Pereira, who was trapped below. Caught in the rush of water from the second explosion, the photographer had drowned.

At the hotel, an hour's drive from town, Steve Sawyer heard the news from Elaine Shaw who telephoned him from her home.

April 22 , 1985
At a meeting of the Convention on International Trade in Endangered Species of Wild Fauna and Flora (CITES), Greenpeace and the Centre for Environmental Education (CEE) successfully opposed five proposals for the creation of turtle farms.

Destruction *The wash-room of the* Rainbow Warrior *is in turmoil after the powerful explosions.*

"It was 1.07 a.m., an hour after the actual bombing," Sawyer says, "when I was called to the phone. A near hysterical Elaine said 'Steve, there's been an explosion and a fire on board the ship, and the *Warrior's* sunk at the dock.' We ran out of the hotel, piled into the car and took off for Auckland. At the police station, the crew was sitting around, totally in shock. Willcox and Edward were getting grilled by the cops. Chris Robinson, who's one of the strongest individuals I've ever known, was close to tears. 'We lost the *Warrior*,' he said to me."

All night the crew sat around, mourning the death of their friend Pereira and the loss of the beloved *Warrior*. The loss of life could have been far greater but, fortunately, many of the people who would normally have been asleep on board were either awake at the time of the blast or were away from the boat.

Sawyer called Maureen Falloon, head of Greenpeace's marine division, and she began contacting the rest of the organization's key people. Patrick Moore immediately telephoned David McTaggart, who was attending an IWC session in England with board members Peter Wilkinson and Monika Griefahn. McTaggart's first thought was that the French were somehow involved, but then he discounted the possibility. "I thought they couldn't be that stupid."

TELEX
ATTN: GREENPEACE ALL OFFICES
10 JULY 1985
FROM: GREENPEACE INT

URGENT   URGENT   URGENT
APPROXIMATELY TWO HOURS AGO,
RAINBOW WARRIOR SUNK BY TWO
EXPLOSIONS   IN   AUCKLAND
HARBOUR,   NEW   ZEALAND.
SABOTAGE SUSPECTED. VERY LITTLE
NEWS FORTHCOMING AT PRESENT.
ONE CREW MEMBER MISSING.
PLEASE DO NOT, REPEAT NOT, CALL
AUCKLAND OFFICE, AS TELEPHONES
ARE JAMMED. WILL HAVE MORE
INFO IN AN HOUR OR SO.

The Bitter Aftermath *Once the Rainbow Warrior is raised from the harbour bed and placed in dry dock, the full extent of the damage is apparent. The engine room is wrecked and in complete chaos (left), and there is a gaping hole in the hull (right) – an irony for a vessel that has come to symbolize non-violent protest. Captain Jon Castle sits on the quayside (below), his back to the ship. Every item of equipment that could be salvaged is laid out on the deck, ready to be sold off under the auctioneer's hammer.*

Tribute to Pereira *Greenpeace France pays homage to the man who died in the* Warrior *bombing.*

April 26 , 1985
Greenpeace began an extensive campaign against NL Chemicals of Ghent and Bayer of Antwerp, who were freshly licensed by the Belgian government to dump titanium dioxide waste in the North Sea. Greenpeace activists boarded the NL Chemicals dump ship Falco *on two occasions, and the* Sirius *was later used to blockade the passage of Bayer's dump ship the* Wadsy Tanker *in Antwerp harbour. As a result, Bayer claimed damages against Greenpeace, and the Belgian authorities confiscated the* Sirius *at the beginning of May.*

On June 30 came one of the most daring Greenpeace escapades to date. In only 15 minutes, Sirius *crew members severed the ship's docking chains and sliced off the upper half of her masts to enable her to pass under several low bridges on the Schelde-Rhine Canal, a waterway serving Antwerp. Under cover of darkness, the* Sirius *crossed the Belgian-Dutch border to freedom.*

The Belgian government announced that it would demand an end to the dumping of titanium waste in 1987.

Nowhere was the shock felt more deeply than in the little office in New Zealand itself. "The phones started ringing as soon as we got in the door at seven the next morning," says Elaine Shaw. "And people were asking what they could do, so we gave them donation boxes to take downtown. The slogan was 'You Can't Sink A Rainbow'."

The New Zealand police reacted swiftly to the first act of terrorism committed on their soil, and began an investigation on an unprecedented scale. Soon there were 48 box files of information, the product of almost 100 officers collecting 400 statements and 1,000 exhibits.

A picture of the French agent's activities on the night of the bombing was soon pieced together from scraps of information from members of the public. Passers-by had seen the Zodiac being launched, fishermen had heard its motor as it crossed the harbour and a cyclist had seen the dinghy and heard a splash, which later proved to have been the outboard motor being jettisoned.

Night-watchmen at a boat club had reported the suspicious behaviour of a man in a wetsuit who had been loading gear from an inflatable into a van parked nearby. Fearing that the man had robbed yachts moored in the harbour, they had noted the van's registration number.

The police traced the van to a rental company near Auckland's airport and learned that it had been hired by a foreign couple called Turenge – agents Mafart and Prieur. When the Turenges returned the camper, they were apprehended and taken in for questioning.

THE FRENCH CONNECTION

Instinct told Detective Superintendent Allan Galbraith, who was heading the investigation, that France was somehow connected with the explosions on board the *Rainbow Warrior*. When the Swiss passports of the French-speaking Turenges proved to be false, and a telephone call they made in custody was traced to the French secret service in Paris, he knew he had been right.

Appeals for information from the general public established that the Turenges had met the crew of the yacht *Ouvéa*, which was now moored 1,060 kilometres (660 miles) away at Norfolk Island, preparing to leave for New Caledonia. But when the police searched the yacht and questioned the crew they could find nothing that would give them grounds for holding her. Later, however, forensic scientists discovered minute traces of explosives in samples they had taken during their search of the yacht. But it was too late. The *Ouvéa* had sailed into international waters, and she was never seen again.

David McTaggart was meanwhile following up his own leads. "It became clear to me, and it was confirmed to me later, like a few weeks later, that it was the French government that was responsible. I then took a French translator to go into France to try to track down Maniguet and Bonlieu," recalls the Greenpeace chief. "In an underground scene I did get to Maniguet, who was hiding and afraid of being blown away. They were trying to cover everything up. I worked really hard. I had a big file on every name, every person, and their aliases, and I reviewed it with the New Zealand police. We'd send reporters after certain leads and kept the pressure on."

The French media began their own investigations and, in their editions for Thursday, August 8, two weekly news magazines, *VSD* and *L'Evénement*, accused French agents of the sabotage. President Mitterand ordered an immediate investigation into the affair, to be conducted by former Gaullist official Bernard Tricot.

After a 17-day inquiry Tricot announced that "on the basis of the information available to me at this time, I do not believe there was any French responsibility." French agents had been in New Zealand, he said, but simply to spy on Greenpeace – certainly not to destroy its ship. That assertion was the subject of a feature in *Newsweek*'s international edition. The title said it all: "Whitewash?"

The official French report did include a few new facts, however. It disclosed that in late 1984 or early 1985, Admiral Henri Fages, at that time director of France's nuclear test centre in the Pacific, had discussed

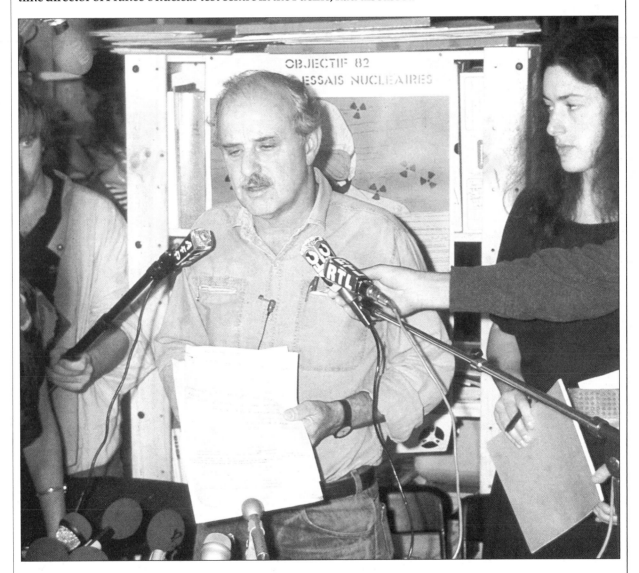

Press Conference *In response to the demand for information, David McTaggart and Monika Griefahn address newspaper and television journalists in the Paris office of Greenpeace.*

MORUROA PEACE CONVOY

Despite the bombing of the *Rainbow Warrior*, Greenpeace decided to continue the protest at Moruroa with its new boat, the MV *Greenpeace*. This boat was equipped and ready for Greenpeace's first expedition to Antarctica, but left earlier than planned in order to complete the *Warrior*'s mission.

Once again Greenpeace would be accompanied on its campaign by a peace flotilla, which this time included: the *Vega*, skippered by Chris Robinson and with Peter Willcox and Grace O'Sullivan from the *Rainbow Warrior* on board; the *Alliance*, a 13-metre (42-foot) scow; the *Varangian*, a 10-metre (33-foot) double-ended cutter; and the *Breeze*, a 24-metre (80-foot) brigantine.

Several boats in the flotilla were sponsored by peace groups in New Zealand, and a full-page newspaper advertisement brought in offers of food, supplies and equipment as well as willing hands to prepare the boats for the trip. Members of the Engineers and Boilermakers Union helped to do engine repairs on the *Alliance*, and the *Breeze* had her rigging overhauled and extra weight added to her keel to give her greater stability.

The MV *Greenpeace* passed through the Panama Canal on her way to lead the protest, and on September 18 spotted a French warship on her radar. Skipper Jon Castle wired the French Defence Ministry: "No need to shadow us. Our mission is a peaceful one." This same French vessel later prevented a crew from the French news agency Gamma rendezvousing with the protest ship at the Marquesas Islands in northwest Polynesia.

Threatening Behaviour

The French placed three warships in the area, carrying specially trained marines to board the protest ships. Although France hoped to avoid a confrontation, the warships had instructions to ram the peace boats if necessary, and had experimented with ways of fouling the propeller of the *Greenpeace*. As the peace flotilla approached the area, the French announced that the exclusion zone around Moruroa had been extended from 19 to 48 kilometres (12 to 30 miles).

The mood aboard the MV *Greenpeace* was a determined one. As Gerd Leipold recalls, "We knew Fernando Pereira quite well. There was tension aboard the ship. There

was huge interest in the whole voyage. The reporters expected a war. I told them this was not what we were after. But France tried everything to prevent us going and to keep away the publicity.

"Essentially we had a peace fleet, and what made the French especially nervous is that we had satellite transmission and a good telephone link and the capability of sending photographs. Besides, a big ship like *Greenpeace* gives you the possibility of staying there a long time. When we were just outside the

---

MV Greenpeace *Built in 1959, the new flagship started life as an ocean-going tug and salvage vessel owned and operated by Smit International. In 1977 the company sold the ship to the Association of Maryland Pilots in the USA, and she was converted and operated as a Pilot Station vessel until 1985, when the AMP became a shore-based operation and no longer needed the ship. When they heard of Greenpeace's plans to sail to Antarctica they donated the 58-metre (190-foot) vessel as a show of support. Her refit provided her with an ice-strengthened hull, a crane, satellite communications and, in 1986, a helicopter landing pad.*

"Moruroa Mon Amour" Gerd Leipold, the Nuclear Free Seas campaign coordinator, stands on deck as a French warship steams past.

12-mile zone, cruising around Moruroa, the *Vega* with us, two French warships followed us."

The crew built a small flotsam raft, which they decorated with a Greenpeace banner and floated around the convoy, but they did not want the *Greenpeace* herself to enter the territorial zone and risk arrest as she was loaded with equipment for the imminent trip to Antarctica.

As it turned out, on October 17, the *Greenpeace* was forced to head to Papeete for repairs when one of her electricity generators broke down. The French authorities there refused to allow her to enter their territorial waters and the ship had to return to New Zealand.

Last Attempt

The following week, the underground nuclear test at Moruroa was set to go. As if to defy world opinion, Prime Minister Laurent Fabius and Defence Minister Paul Quiles arrived to watch it. Greenpeace made a last-ditch effort to halt the blast, sending the *Vega* into territorial waters within 6.5 kilometres (4 miles) of Moruroa, but her protest voyage came to an end when eight French commandos boarded the yacht and arrested the crew.

Under Full Sail The *Breeze and her crew head for Moruroa to join the peace convoy.*

May 14 , 1985
Greenpeace protesters unfurled a banner reading, "Put Words Into Action – Stop Nuclear Testing", above the Soviet Embassy in Vienna during the sixth session of the Conference on Disarmament in Europe.

June 5, 1985
In New Zealand, Greenpeace activists chained themselves to the main gate of Ivon Watkins-Dow Ltd in New Plymouth, the last factory in the world still making the extremely hazardous herbicide 2,4,5-T.

June 11, 1985
Greenpeace launched its European campaign to lobby the EEC for a ban on the import of kangaroo products.

July 9, 1985
To coincide with meetings in Helsinki of the Geneva Convention on long-range transboundary air pollution in Europe, and to protest against the UK's tardiness in agreeing new pollution control targets, three Greenpeace protesters – Milo Dahlmann, Birgit Seffmark and Paulette Agnew – scaled a 116-metre (380-foot) construction crane next to one of the vast cooling towers at the Drax power station in Yorkshire. None of the towers are fitted with any cleaning devices.

It was estimated that by 1986 the station would be emitting about four million tonnes of sulphur dioxide annually, more than the whole of Scandinavian industry.

Greenpeace's proposed protest voyage to Moruroa with Admiral Pierre Lacoste, head of the DGSE. He was especially concerned about reports that Polynesians would be joining the protest. On March 1, Fages had sent a note to Defence Minister Charles Hernu calling for heightened efforts to "forecast and anticipate the actions of Greenpeace". The word "anticipate" – which in French is synonymous with "ward off" and "prevent" – was obviously taken literally.

BREAKING THE COVER

Despite repeated denials by the government that they had played any part in the sabotage of the *Rainbow Warrior*, leading French newspapers insisted that this was simply part of a cover-up, and they demanded that the truth be revealed. Then, on September 17, *Le Monde* declared that there was now no doubt that Charles Hernu and Admiral Lacoste had been aware of the operation - indeed, it seemed likely that it had been carried out under their orders.

"President Hernu. Official Portrait" This cartoon by Guiraud originally appeared in the French satirical magazine Canard Enchainé *on November 6, 1985.*

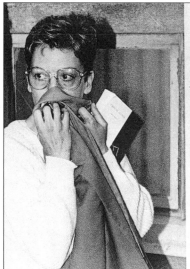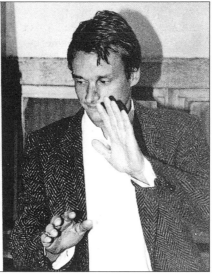

Exposed *Captain Dominique Prieur and Major Alain Mafart of the French DGSE attempt to shield their faces from the cameras.*

Just 48 hours later, Admiral Lacoste was dismissed and Defence Minister Hernu resigned. For a time it seemed that President Mitterand himself would have to resign. Then, on the evening of Sunday, September 22, Prime Minister Laurent Fabius appeared on television to make an announcement to the nation and the world: "Agents of the DGSE sank this boat. They acted on orders. This truth was hidden from state counsellor Tricot."

Six weeks later, on November 4, 1985, the preliminary hearing in the trial of agents Mafart and Prieur began in Auckland. The hearing was expected to last for weeks. More than 100 witnesses were to be brought by the prosecution and 147 reporters, photographers and film crews from around the world were there to record the action.

ROUGH JUSTICE

But there was to be no long, dramatic murder trial. The police evidence was not strong enough to prove murder and arson, nor that the two agents had been responsible for the actual placing of the bombs. As a result a deal was struck before the agents got into court. Mafart and Prieur pleaded guilty to charges of manslaughter and wilful damage, and the trial was over in just 34 minutes. They were later sentenced to 10 years on the first charge and seven years on the second, the sentences to run concurrently.

Some kind of justice had been seen to be done, but the story was far from over. The French authorities were certainly relieved that a lengthy trial, at which a parade of damaging evidence would have been presented, had been avoided. Yet the cynical motives of the mission had exposed them to international ridicule and humiliation, and had achieved the exact opposite of its presumed objective.

As for Greenpeace, the tragic death of Pereira and the loss of the *Warrior* seemed to underline the seriousness of its endeavours and to heighten its sense of purpose. Far from being deterred or defeated, Greenpeace was, in the next few years, to spread its name, influence and activities across a wider landscape than ever before.

SOUTH AFRICA'S ECONOMIC TREMORS

Newsweek
THE INTERNATIONAL NEWSMAGAZINE
September 9, 1985

WHITEWASH?
France's Greenpeace Report

NUCLEAR FREE PACIFIC

GREENPEACE

HERNU NE SE
REND PAS

DÜNYA

YEŞİL BARIŞ HAREKATI

"Biz dünya vatandaşıyız"

Bundan 15 yıl önce 12 adamın nükleer denemelere karşı başlattığı pasif direniş,
bugün 1,5 milyona yakın üye ve sempatizanı bulunan "Greenpeace" (Yeşil B...
örgütü tarafından sürdürülüyor.

scandal French fias

THE GUARDIA
Printed in London and Manchester
Monday September 23 19

Lange denounces 'sordid act of
terrorism' after Fabius admission

France
admits
ordering
sinking

From Campbell Page
in Paris

France admitted yesterday
that it ordered military frogmen to sink the Greenpeace
flagship, the Rainbow Warrior, in Auckland Harbour,
more than two months ago.

An angry New Zealand Prime
Minister, Mr David Lange, last
night said that the attack was
"a sordid act of international
state-backed terrorism."

New Zealand's position in the
affair had been vindicated by
the French admission.

Mr Lange was annoyed that
the French Prime Minister, Mr
Laurent Fabius, would not name
the French agents and exonera-
ted them from blame.

In his statement, Mr Fabius
said that the government would
look for those who had given

**Laurent Fabius ...
the cruel truth**

least a short breathing space
before further announcements.

The resignation of the Defence Minister, Mr Charles
Hernu, and the sacking of the
DGSE chief, Admiral Pierre
Lacoste, on Friday were
thought to be as much of a
declaration of intent as could
be expected until Mr Quiles
had a little time to pursue and
evaluate his inquiry.

Mr Fabius's admission came
only a week after President
Mitterrand told journalists that
French operations in New Zealand had been legitimate, and

the orders and not those who
had obeyed them. The latter had
carried out instructions, and in
the past had fulfilled dangerous
missions for their country.

The Prime Minister, flanked
new Defence Minister,
referred last
estab-

Leader comment, page 12

nokta

DES TE

lepoin
HEBDOMADAIRE D'INFORMATION / 6 OCT. 1985

GREENPEACE
jusqu'où?

**Amiral Pierre Lacoste,
Charles Hernu,
Laurent Fabius
et François Mitterrand**

HOME EDITION

French pair
get 10 years
for role
in bombing

Front-Page News
The story of the
Rainbow Warrior
bombing made
headlines around
the world.

SPY BLEW U

PEACE SHIP

l'Unità **OGGI**

VENERDÌ
23 AGOSTO 1985

Uomini della Dgse dietro l'affondamento del Rainbow Warrior

L'attentatrice è un capitano

Ormai è provato: i servizi
francesi contro Greenpeace

In carcere in Nuova Zelanda c'è anche un maggiore dei sommozzatori di stanza in Corsica - Identificata la «talpa» infiltrata tra i pacifisti: agente pure lei, col grado di tenente

Weather:
Fair

HOME EDITION

Sab

BLA

El premier francés responsabilizó a ... el sabotaje a Greenpeace

THE WAIKATO TIMES©

HAMILTON, NEW ZEALAND · THURSDAY, JULY 11, 1985 · ❋ FIRST EDITION · Price 30c Rural Delivery Price 35c

Killer blasts sink ship

SHINSEI

NEWSPAPER FOR PEACE AND DEMOCRACY

JUL.15.1985
No.139·140 Y250

1985年7月15日 (10)

抗議船 爆破さる

as French spy scandal

POURQUOI LES GAUCHERS GAGNENT

le nouvel **Observateur**

LE SCANDALE
GREENPEACE
QUI A DONNÉ
L'ORDRE ?

ROPA · i veckan · TEMPUS

Le Monde avslöjar lögnerna
...ng attentatet mot Greenpeace

...ursday, July 11, 1985

Auckland Star

...land's largest evening newspaper — Inquiries 797-626, Star classifieds 794-666.

Price 30c (Auckland area home delivery: $1.60 a week)

...ge, says Greenpeace

'spied on Greenpeace'

CHAPTER TWELVE

GOING GLOBAL

A CHRONOLOGY OF GREENPEACE'S
MOST RECENT ACTIONS

"GREAT ARE THE TASKS AHEAD, terrifying are the mountains of ignorance and hate and prejudice, but the Warriors of the Rainbow shall rise as on the wings of the eagle to surmount all difficulties. They will be happy to find that there are now millions of people all over the earth ready and eager to rise and join them in conquering all barriers that bar the way to a new and glorious world. We have had enough now of talk. Let there be deeds."

Warriors of the Rainbow, *Willoya and Brown*

1985

July 15

The *Sirius* arrived in Bournemouth, England, at the start of the IWC meeting there, to deliver one million signatures demanding a moratorium on whaling.

July 29

Seven Greenpeace cyclists let down the tyres of a truck carrying radioactive waste across the Columbia River Bridge at Plymouth, Washington, and were arrested.

August 9

Sirius arrived in Iceland to campaign against "scientific" whaling. The government was allowing the killing of 200 whales each year for the next four years, the meat and oil to be exported to Japan. The Greenpeace delegation met with the Icelandic Fisheries Minister, and held public meetings in an attempt to stop this.

August 28

Greenpeace's hot-air balloon *Trinity*, held since the 1983 flight over the Berlin Wall, was released by the East German authorities.

September 19

A Greenpeace representative met with Prime Minister Atli Dam of the Faroe Islands to protest about the annual hunt of pilot whales.

September 21

In Darwin, Australia, Greenpeace divers delayed the docking of the *Clydebank* for five hours. The ship was due to pick up 42 containers of "yellow cake" from the Ranger uranium mine.

November

Greenpeace Canada continued its campaign to end driftnet fishing on the high seas. Protesters symbolically entangled the Japanese consulate in Vancouver, and petitioned the Japanese government, who had banned the use of drift nets in its own waters, to phase out its driftnet fisheries completely.

September 21 A Greenpeace diver appears in front of the Clydebank *off Darwin, Australia, in an attempt to stop the ocean transportation of uranium.*

GREENPEACE BELGIUM

At the end of 1985, Greenpeace Belgium was brought wholly under the jurisdiction of the Greenpeace Council, but its roots can be traced back to the early 1980s.

In 1983, a one-man office was opened in Brussels and, at the end of 1984, ten people were hired for one year to coordinate campaigns relating to EEC policy, acid rain, Antarctica and toxic chemicals.

Until 1986, Greenpeace Belgium was essentially providing support for campaigns that originated elsewhere but that year, after a change of office and personnel, the first truly Belgian campaign took off with the arrival of the *Beluga* for a month-long tour. Late in 1986 Greenpeace successfully delayed the ocean incineration vessel *Vulcanus II* from leaving the port of Antwerp. In 1987 the *Beluga* returned for an extensive tour of Belgium, and in October that year the boat was seriously damaged when it blocked the *Vulcanus II* in one of the Antwerp locks.

WATER FOR LIFE

The aim of Greenpeace's four-month "Water For Life" campaign was to push for a comprehensive plan to clean up the Great Lakes.

In recent years, the lakes, the largest freshwater reservoir in the world, have become increasingly contaminated due to the amount of pesticides, polychlorinated biphenyls (PCBs), dioxin and other chemical waste pouring into them from industries around the Great Lakes basin.

A report by the US National Research Council and the Royal Society of Canada revealed that the 37 million people living around the Great Lakes Basin have "appreciably more" toxic chemicals in their bodies than other North Americans.

For this campaign, Greenpeace leased the *Fri*, which called at Great Lakes ports from Quebec to Wisconsin, focusing on known toxic "hotspots" including Niagara Falls, Sarnia, Toronto, Chicago and Midland, Michigan.

The campaigners combined a busy schedule of meetings with media events and direct actions aimed at maximizing attention for the issue. They lobbied chemical companies and met with public interest groups such as the Grand Calumet Task Force, whose aim is to clean up the Grand Calumet River, which drains into Lake Michigan.

Stopping the Flow

In Midland, Michigan, worldwide corporate headquarters of the Dow Chemical Company, the sixth largest chemicals conglomerate in the world, Greenpeace members used homemade wooden plugs, painted red with the word STOP in white lettering, to block the company's main effluent pipes, which spew 90,000 litres (20,000 gallons) of chemical waste a day into the Tittabawassee River and down into Lake Huron. It took the company three days to

***Clean-up Squad** The* Fri *sails up the Fox River, Wisconsin.*

remove the stoppers and, on July 15, Melissa Ortquist and her two male companions returned to put the plug back in. Before this could be achieved police frogmen swam over and arrested them.

Two other protesters were later arrested when they climbed two 43-metre (140-foot) towers at a Dow plant and attempted to string a banner between them that read: "You Don't Know What You Got Till It's Gone." All five were charged with trespass and spent three days in jail before Greenpeace could find the $30,000 bail needed to release them.

The Greenpeace action made headline news, as did the attempt by a Dow official to discredit Melissa Ortquist – and Greenpeace itself. This official publicized the results of a routine blood test that had shown the campaigner to have venereal disease. Subsequent tests proved the result to be false, and as a result the US president of the company later wrote to Melissa Ortquist apologizing for the company's "serious error of judgement".

In addition to this action, Greenpeace members also:
• Camped for a weekend on the highest point of the Blue Water Bridge, linking Sarnia and Port

Huron, to protest against toxic dumping. The area around Sarnia is a "chemical valley" in which there are 11 industries involved in petrochemicals and oil refining discharging contaminated water directly into the St Clair River.
• Staged protests at Green Bay, Wisconsin, where the Fort Howard Paper Company discharges PCBs and other toxic pollutants into the lower Fox River.
• Plugged a 1-metre (3-foot) diameter sewer pipe discharging toxic waste from three chemical plants into a stream called Fields Brook in Ashtabula, Ohio. After two hours Greenpeace itself removed the plug – a first for the organization – when company officials claimed that the backflow it created would permanently damage the plant.
• Drew attention to the pollution of the Niagara River, which forms the boundary between the State of New York and the Province of Ontario. The river area is a major centre of chemical production, and the Occidental Chemical Company's plant, which occupies 40 hectares (100 acres) on the bank of the river, discharges about two tonnes of chlorinated organic poisons into the Niagara every month.

"Clean Water Now" A Greenpeace stall dramatizes the dangers of polluted water.

GREENPEACE SWEDEN

At the Greenpeace Council's AGM, Greenpeace Sweden became a self-sufficient national office with voting status for the first time.

Greenpeace's presence there had begun with a young woman named Lena Hagelin, who had initially worked at the office in Paris. Her role in the protest at the Swedish Embassy there, against the sending of Swedish nuclear waste to France for reprocessing, made headlines in her homeland. She returned to start a Greenpeace office in the spring of 1983, with Janus Hillgaard from Denmark.

That autumn Goeran Olenborg, later to become chairman, was hired and they were joined by toxics campaigner Haakan Nordin in 1984 and ocean-ecology campaigner Jakob Lagercrantz in 1985. By 1986, there were 10 full-time people working, rising to 15 in 1987 and 30 in 1988.

The first office was a single room in Göteborg, but the rapid growth of the organization required them to move into ever larger premises. At the end of 1985 a branch was opened in Stockholm.

Membership at the end of 1983 was 1,500, which had risen to 162,000 by April 1988. Total income was SEK 23 million in 1987.

"Stop Acid Rain" A Swedish campaign badge.

ANTARCTICA – THE FIRST ATTEMPT

Greenpeace's campaign to conserve Antarctica as a "World Park" is guided by four main principles:
•Antarctica's value as a wilderness should be protected.
•There should be complete protection for wildlife (though limited fishing would be permissible).
•Antarctica should remain a zone of limited scientific activity, with co-ordination between all nations.
•Antarctica should remain a zone of peace, free of all weapons.

Greenpeace believes that there is no intrinsic need to establish a new legal entity to implement these principles. They could be enacted either through the Antarctic Treaty System or through the mechanisms of the World Heritage Convention. This lays down four criteria, any one of which qualifies an area to be designated as a World Heritage Site. Antarctica fulfils all four.

As Greenpeace's campaign developed, it became clear that a base would have to be set up in Antarctica to: monitor the activities of the nations already established there; carry out scientific research; and draw public attention to the future of the continent. Such a base would also potentially give access to the formal meetings of the Antarctic Treaty members.

By the end of 1985, after months of preparation, the expedition ship MV *Greenpeace* set sail loaded with crew and supplies, a specially designed base hut and a helicopter to ferry the cargo ashore. However, by the time they reached Antarctica, sea-ice conditions were the worst for 30 years and, after numerous attempts to break through to the coast of Ross Island, the campaigners had no choice but to return to New Zealand and reconsider their plans.

Ice Watch A fish-eye view from the crow's nest of the MV Greenpeace *as she sails through Antarctic waters.*

1986

February 27

A sailor on board the Japanese whaling ship *Shouan Maru* threatened a Greenpeace protester with a knife after she chained herself to the railing of the vessel.

March 31

A three-week tour of Irish sea ports by the *Sirius* culminated in a series of "soft" actions off the Cumbrian coast designed to highlight the dangers of radioactive waste discharges from Sellafield.

Some 3,500 balloons, each bearing a radiation symbol, were released from the boat to symbolize the daily airborne releases of radioactive gases from the plant.

The following day, designated British Nuclear Fool's Day, 1,000 wooden triangles bearing the legend "Released Off The Windscale Pipeline, April 1986" were placed in the sea.

On April 4, in Barrow docks, *Sirius* crew members joined local activists in draping banners and painting slogans on two ships, the *Mediterranean Shearwater* and the *Pacific Teal*, used to transport spent nuclear fuel from Europe and Japan to Sellafield for reprocessing.

April 4

Four Greenpeace teams, 14 people in all, set out from different locations to invade the Nevada test site in an attempt to hinder a US test, code-named Mighty Oak, that risked provoking the USSR into abandoning their moratorium on nuclear testing. The presence of Greenpeace delayed the test for two days. When it came, the explosion accidentally irradiated millions of dollars' worth of equipment. The USSR declared that, despite this action, they would extend their moratorium until August 6.

On the same day, Greenpeace activists plugged a waste pipe from the Monsanto herbicide production plant in Antwerp, Belgium, by bolting a metal shield onto the outlet and covering it with sandbags.

The Monsanto discharge is a particularly striking example of environmental mismanagement as not only does the pipe discharge into the Scheldt River, but the outlet is situated in a bird sanctuary – the Galgenschoor nature reserve – which is a legally-protected wetland.

April 17

Greenpeace UK launched its "Shut It" campaign, which set out a 10-point plan and a two-year deadline for ending nuclear reprocessing at Sellafield. The campaign was designed to coincide with a meeting of the Scientific Committee of the Paris Commission, the international body that regulates discharges of all toxic substances from pipelines into rivers and coastal waters.

May 15

Greenpeace Sweden began its long-term campaign against pollution caused by the pulp and paper industry by demonstrating against the Vaeroe Bruk company who routinely discharged chlorinated waste directly into the waters of the Kattegatt. Five hundred kilograms (1,100 pounds) of damaged fish, caught in the areas around the factory's outlets, were dumped outside the gates as a "gift" to the management. Local fishermen and labour unions lent their support.

April 4 Unfurling a flag and banner at the Nevada test site.

May 30

A crew of 11 Greenpeace activists from six countries set sail on board the MV *Moby Dick*, heading for a confrontation with the Norwegian whalers in the northeast Atlantic. The ship, a 120-tonne, 25-metre (82-foot) former Dutch fishing vessel built in 1959, had been converted into a Greenpeace campaign ship in just over two weeks.

Twelve days later, the *Moby Dick* and her crew were under arrest in the northern Norwegian port of Vardo. They had been seized by the coastguard and charged with trespass after stopping the whale hunt for more than eight hours.

After eight days, both ship and crew were released, and the search for the whalers began again. On June

June 26 Greenpeace blocks the waste pipe of Neusiedler AG's pulp and paper mill in Austria (right).

July 14 Crew members from the Moby Dick *occupy the crow's nest of a Norwegian whaling ship (below).*

24, they again disrupted the hunt, and were again arrested, this time being charged with "handling explosives in a dangerous way without a licence". The protesters had tied a knot in the cable of the explosive harpoon.

The next day, local police stormed on board and seized the inflatables. It was another two weeks before the *Moby Dick* was free again. The last protest of the campaign came on July 14, when two crew members managed to sneak aboard a whaler refuelling in port. They occupied the crow's nest and draped a banner, staying seven hours before being hauled down and arrested.

When the US government finally threatened sanctions, Norway announced that they would cease whaling after 1987, but they were later to use the "scientific whaling" loophole to continue the hunt.

In August 1986, the captain of the *Moby Dick*, Jon Castle, agreed to pay fines totalling £2,740 for boarding Norwegian whaling vessels.

June

More than a dozen Swiss Greenpeace demonstrators chained themselves to trees at the resort of Crans-Montana in the Valais, venue for the 1987 World Ski Championships, to protest against the felling of 50,000 square metres (60,000 square yards) of forest to provide facilities for the event.

June 3

The *Vega* set off on a "Pacific Peace Voyage" to focus attention on the drawbacks of the Rarotonga (Nuclear Free Zone) Treaty and investigate possible ocean ecology problems in the Pacific.

June 26

Activists from Greenpeace Austria blocked waste water pipes from Neusiedler AG, one of the worst polluting pulp and paper mills in the country. The company's fire brigade used their hoses to try to prevent the action.

SIRIUS IN THE MEDITERRANEAN

During the summer of 1986, the Sirius *set out from Sevilla on a campaign tour covering many issues that affect the Mediterranean ecosystem.*

The first action came on May 17, when inflatables from the *Sirius* intercepted the transport vessel *Mediterranean Shearwater* as it passed through the Straits of Gibraltar. Crew members boarded the vessel and occupied a crane. Bound for the reprocessing plant at Sellafield, the vessel was loaded with 30 tonnes of spent nuclear fuel from the Italian reactor at Latina. The ship was boarded again some months later by *Sirius* crew members near the Italian harbour of Civitavecchia.

In between these two actions, the *Sirius* put into the coastal town of Palomares where, on January 16, 1966, four nuclear weapons fell after a collision between two US military planes. Two of the bombs broke up and contaminated the area with plutonium. Greenpeace's presence lent strength to the local people's court appeal for compensation.

The "Italian Bar"

The second stage of the campaign was related to over-fishing. Greenpeace protesters confronted coral boats in the Sea of Alboran, between Spain and Morocco, which were using a particularly destructive means of harvesting red coral from the sea-bed. Known as "the Italian Bar", this consists of an iron rail dragged by trawlers across the sea-floor. As a result of Greenpeace's actions this practice was banned that September.

In a related action, to dramatize the illegal use of fishing nets in coastal areas, Greenpeace built and sank a number of artificial reefs, made out of steel drums and iron bars, which bore the legend: "If Your Nets Have Become Entangled With This Artificial Reef, It Is Because You Are Fishing In A Prohibited Area".

Mud Bath *Lloyd Anderson is sprayed with toxic sludge during the blocking of the Portman Bay pipe.*

The *Sirius* sailed on to the main island of the Cabrera archipelago, part of the Balearic Islands, an area of great natural beauty. Greenpeace was there to add weight to demands for an "environmental impact" study of the military exercises that had been taking place on Cabrera annually since 1973. The planned manoeuvres that month were abandoned as a result, and the Ministry of Defence agreed to postpone further exercises until such a study was completed. Greenpeace was to return in September that year to help the local environmental group block another military mission.

The *Sirius*'s next port of call was another threatened archipelago, the volcanic islands of Columbretes. The crew collected a tonne of garbage there to highlight the damage to the area, and this was later dumped in front of the headquarters of the Regional Government of Valencia.

The final actions of the summer were directed against toxic pollution. On July 31, Greenpeace volunteers chained themselves to a pipe discharging toxic sludge from a lead and zinc mining operation, owned by Peñarroya España SA, into the bay of Portman in Murcia, and partially blocked it. Some 7,000 tonnes of sludge have poured into the bay every day for almost three decades.

Then, on August 13, the *Sirius* intercepted two Spanish ships, the *Nerva* and the *Niebla*, in the Gulf of Cadiz, where they were preparing to dump their cargo of highly toxic waste. Two Greenpeace volunteers occupied the ships' discharge pipes, preventing the dump taking place for 20 hours. The following night, they again boarded the ships and repeated their success.

On August 21, the *Sirius* left the Spanish coast to make the return journey to Amsterdam.

Portman Bay *Greenpeace activists chain themselves to the cage that surrounds the toxic waste pipe.*

July 8

The two convicted *Rainbow Warrior* saboteurs, Major Alain Mafart and Captain Dominique Prieur, were released from jail in New Zealand for three years' "confinement" on the French military atoll of Hao. This was part of a UN-mediated deal between France and New Zealand that also included an official French apology to the New Zealand government for the attack on the *Warrior*, payment of nearly $5 million in compensation, and an end to French obstruction of New Zealand imports. Prime Minister David Lange announced later in the year that part of the compensation money would be used to improve the equipment that monitors French nuclear tests in the Pacific.

August 13

Two Greenpeace climbers scaled a 61-metre (200-foot) smokestack at the Stauffer Chemical Company facility in Hammond, Indiana, dramatizing local opposition to the chemical company's plans to build a hazardous waste incinerator. Health authorities later refused permission for the construction of the plant.

September 25

Following the Chernobyl disaster in April, Greenpeace staged an action at the famous Big Wheel in Vienna to coincide with a Reactor Safety Conference held by the International Atomic Energy Agency (IAEA) in the city. The group also presented a report, compiled by an international panel of experts, that reviewed the safety of every commercial power reactor currently operating and concluded that no reactor design can be considered safe.

October 2

Two Greenpeace protesters climbed to the top of the Sydney Opera House, where they hung a banner calling for an end to French nuclear testing in the Pacific.

October 12

During the summit meeting in Iceland between Ronald Reagan and Mikhail Gorbachev, the *Sirius* tried to sail into Reykjavik harbour to display a banner reading "The World Demands A Test Ban Treaty". But as the crew unfurled the banner, the *Sirius* was rammed by an Icelandic warship and boarded by a dozen coastguards, who arrested the crew and escorted the ship to a fishing port south of the capital.

November

The *Vega* arrived in Darwin, Australia, to protest against the export of a consignment of 800 tonnes of "yellow cake" from the Ranger uranium mine to the US and Canada. She prevented the *Forthbank* from mooring for almost an hour and, as a result, the *Vega*'s skipper, Chris Bone, was arrested. A Greenpeace protester who chained herself to a crane 30 metres (100 feet) above the ground was arrested, as were four other demonstrators.

November 10

In Greenpeace UK's biggest operation on land, 30 activists from five countries invaded the site of the Ferrybridge coal-fired power station in south Yorkshire to highlight the dangers of acid rain produced by emissions from power stations.

CAMPAIGN CONTROVERSY

One of the most contentious campaigns in the recent history of Greenpeace was the anti-fur campaign launched by Greenpeace UK in September 1984, using a powerful poster and cinema advert by the photographer David Bailey.

The campaign was designed to highlight the cruelties of the leg-hold trap, which is still used to catch the majority of wild fur-bearing animals for the trade, and to dissuade potential consumers from wearing fur.

The power of Bailey's imagery guaranteed a high public profile for the campaign, but it was soon clear that it was causing considerable problems within Greenpeace. The offices in Canada and Denmark had developed working relationships with the Inuit peoples, who depended on the fur trade for their livelihood. After long deliberations, the Greenpeace International council voted to end the fur campaign.

"It takes up to forty dumb animals to make a fur coat. But only one to wear it."

November 13

Following six months of intensive campaigning by Greenpeace, eight multinational sports shoe companies – Nike, New Balance, Lotto, Tacchini, Adidas, Puma, Diadora and Mitre Sports – pledged to stop selling kangaroo-skin shoes in the UK.

November 17

Costumed Greenpeace activists in Hamburg demonstrated outside a store selling reptile products, in a symbolic protest against the trade in endangered species.

November 21

Greenpeace protesters obstructed the entrance to the Wackersdorf construction site in West Germany, where a nuclear reprocessing complex was being built.

December 8

Greenpeace protester Craig Maddock was threatened with tear gas and Mace when police came to remove him from the crow's nest of the Antarctic supply ship *Polarbjørn* in Hobart, Tasmania. Two other members, Christian Bell and Laurie Newman, were also arrested when they chained themselves to earth-moving vehicles on the dock. Greenpeace was campaigning to stop the loading of equipment for use in the construction of the French airstrip in Antarctica.

December 16

Greenpeace climbers formed a "living curtain" beneath a bridge over the Rhine near Leverkusen, West Germany, in order to block shipping while divers took samples from chemical plant discharge pipes in mid-river. They also hung a banner reading "Stoppt Die Brunnenvergifter – Wasser Ist Leben" (Stop The Poisoners – Water Is Life).

December 19

Four Greenpeace climbers scaled a 120-metre (400-foot) chimney at the

December 16 Greenpeace climbers stop Rhine traffic with a "living curtain" while samples are taken.

Basel works of the Ciba-Geigy chemical company in Switzerland, in an anti-pollution protest aimed at drawing attention to Greenpeace demands for stricter controls in the chemical industry.

The action was timed to coincide with a conference in Rotterdam at which ministers from the Rhine states met to discuss one of the worst environmental disasters of the decade.

Several weeks earlier, a fire at the Sandoz chemical plant in Basel had resulted in 30 tonnes of pesticides, fungicides, herbicides and other toxic substances pouring into the Rhine. Consequently, over 300 kilometres (190 miles) of the Rhine died. A 60-kilometre (37-mile) slick of toxic chemicals wound its way down-stream, poisoning at least half a million fish and more than 150,000 eels in its path.

ANTARCTICA – THE SECOND VOYAGE

Having failed to set up its "World Park" base in Antarctica in 1985/86, Greenpeace returned for a second attempt in the southern summer of 1986/87.

Once again, the MV *Greenpeace* set out from New Zealand carrying a full crew, the prefabricated base building, supplies for the 12-month stay, and the four winterers. This time they had a larger helicopter with a greater range in case they encountered similar sea-ice conditions to those of the year before.

However, when they arrived in the vicinity of Ross Island on January 25, the Greenpeace flagship was able to moor just 200 metres (660 feet) off the beach at Cape Evans.

Using the helicopter to lift sections of the base ashore, the camp was set up and fully operational within three weeks. In addition to individual sleeping quarters, a communal living area, bathroom and shower, the base contains laboratory facilities, communication equipment and a hydroponic "greenhouse" to provide a supply of fresh vegetables.

The four winterers were carefully chosen for their skills and personal qualities. The leader of the team was a New Zealander, Kevin Conaglen, an experienced mechanic who had previously worked at New Zealand's Scott Base in Antarctica. Justin Farrelly was the radio operator and technician. West German scientist Gudrun Gaudian monitored pollution from the established bases and studied Antarctic marine life. The base doctor, Cornelius van Dorp, carried out research into the effects of low light levels and isolation, as well as looking after the health of the Greenpeace team.

The years of planning and effort had finally paid off. Greenpeace's campaign to save Antarctica had taken a major step forward.

Chilling Task *A smiling Gudrun Gaudian takes samples from an Antarctic base outflow pipe.*

1987

February 5

Eighteen hundred demonstrators, including Martin Sheen, Kris Kristofferson, Carl Sagan and Daniel Ellsberg, assembled at the gates of the Nevada test site in protest at a crucial US nuclear test that the Soviet Union claimed would force them to end their unilateral moratorium. The Greenpeace balloon *Trinity* sailed into the test zone and a number of Greenpeace members trespassed on the site, where they were immediately arrested. In fact, the US Department of Energy had conducted the test, code-named Hazebrook, two days ahead of schedule in order to pre-empt the protesters.

March 9

Greenpeace US activists formed a "human billboard" on the Capitol steps in Washington DC to call for an end to plutonium production. This coincided with another protest on the same day at Hanford nuclear reactor in Washington State, which handles plutonium production for the US Department of Energy.

April A radiation sign marks a nuclear victim's grave in David Bailey's dramatic cinema short "Meltdown".

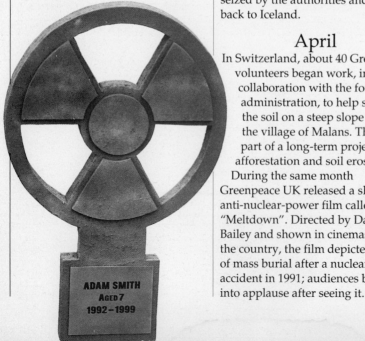

ADAM SMITH
AGED 7
1992–1999

March 22

A shipment of Icelandic whale meat bound for Japan was intercepted by Greenpeace campaigners in Hamburg, West Germany, and the police and customs were informed. The whale meat was labelled as "frozen seafood". Transit of products from animals on the CITES endangered species list is illegal in Germany. The consignment was seized by the authorities and sent back to Iceland.

April

In Switzerland, about 40 Greenpeace volunteers began work, in collaboration with the forest administration, to help stabilize the soil on a steep slope above the village of Malans. This was part of a long-term project on afforestation and soil erosion.

During the same month Greenpeace UK released a short anti-nuclear-power film called "Meltdown". Directed by David Bailey and shown in cinemas across the country, the film depicted scenes of mass burial after a nuclear plant accident in 1991; audiences burst into applause after seeing it.

March 9 "Stop Plutonium Production Now" – a human billboard on Washington's Capitol steps.

April 26

In a spectacular demonstration in the centre of Prague, three Greenpeace protesters from Sweden, Austria and Denmark unfurled a giant banner on the side of the National Museum in Wenceslas Square that read "Chernobyl Never Again – For A Nuclear Free Future". Beneath it, two Austrian Greenpeace members distributed leaflets criticizing Czechoslovakia's nuclear power programme. The leaflets demanded the closure of the country's seven Soviet-built reactors and a stop to two larger plants under construction.

On the same day, Greenpeace protesters in the UK staged a demonstration next to the famous locomotive Flying Scotsman, which was being used by British Nuclear Fuels Ltd to run a sight-seeing tour to the Sellafield plant.

May/June

The *Beluga* carried out a 45-day tour of large areas of the southern Swedish coast and the country's two major lakes, Vaettern and Vaenern. The tour was designed to highlight the problems caused by the untreated discharge of synthetic chemicals, and the main targets were chemical industries, municipal sewage treatment plants and pulp and paper factories.

June 3

Divers attempted – in defiance of a High Court injunction – to block the Sellafield pipelines discharging radioactive waste. The divers cut through the pipes with cutting torches and then inserted two large rubber balls that, when inflated, blocked the pipes.

During the plugging operation, copies of the court injunction were thrown on the deck of the Greenpeace ship *Sirius* from the British Nuclear Fuels tug *Huskiss*.

June 9

Seven Greenpeace members were arrested after the *Vega* was rammed by a police launch at the mouth of the Brisbane River in Australia. The *Vega* had spread an anchor chain across the river in an attempt to stop the USS *Ramsey*, a guided-missile frigate carrying nuclear arms, from entering the harbour. The *Vega* was confiscated by the Queensland Water Police, and the seven Greenpeace crew were still awaiting trial more than a year later, on charges that carried a possible seven-year jail sentence. After negotiations, the *Vega* was released in October.

June 10

BNFL made an application in the High Court for legal costs against Greenpeace. Greenpeace Netherlands and Greenpeace International gave an undertaking that they would not damage or interfere with the pipeline again. BNFL said they would present evidence to show that a previous injunction had been breached and that they would call for the sequestration of Greenpeace assets and the possible imprisonment of Hans Guyt, the campaign leader, and Willem Beekman, skipper of the *Sirius*. They

were also considering a similar action to recover the cost of repairing the pipelines.

June 23

Greenpeace Austria returned a lorry-load of waste, containing high levels of heavy metals, to the company responsible for producing it. Within a few days of the action, the Austrian authorities demanded that the company remove such waste from the banks of the River Ybbs and the River Danube.

July

The *Moby Dick* sailed to Lerwick in the Shetland Isles, Scotland, to take her crew to a conference of parties opposed to the nuclear discharges from the prototype fast reactor (PFR) at Dounreay. The following month the ship returned to Scotland to take samples from the plant's radioactive waste outflow.

July 10

On the second anniversary of the *Rainbow Warrior* bombing, Greenpeace launched a new international initiative against the world's nuclear navies – the Nuclear Free Seas Campaign. The launch of the campaign in the USA was accompanied by spectacular direct actions in Australia, where climbers scaled the highest tower in Sydney, and in Auckland, New Zealand, where Greenpeace activists suspended themselves from the harbour bridge.

July 18

Greenpeace Switzerland drew attention to the traffic problem in the canton of Uri by suspending a 100-metre (330-foot) banner across the

July 18 Sylvia Parli and Renato Ruf hang a banner in Uri.

motorway, reading "Es Reicht" (That's Enough). Greenpeace roller-skaters also distributed a "jam newspaper" to the waiting drivers. Since the opening of the St Gotthard road tunnel, lorry traffic alone in the canton has risen to 2,000 per day. A Greenpeace study showed that on one holiday weekend the traffic – 80 per cent of which was foreign – spewed the equivalent of 30 tonnes of nitrogen oxide, 28 tonnes of hydrocarbons and 63.5 kilogrammes (140 pounds) of lead into the air. About 60 per cent of the canton's forests are now sick or dying.

July 29

Greenpeace member Shelley Stewart offered a public hearing in Seattle the chance to see the mayor's own personal garbage, after a raid on his trash cans a few nights before. Mayor Charles Royer was recommending that a garbage incinerator be set up in the south end of the city, a plan opposed by Greenpeace because incineration contributes to acid rain and smog,

and creates hazardous ash that must be buried. (This secondary waste is so toxic that in most states in the US it is illegal even to bury it.)

Greenpeace claimed that a combination of recycling, reuse of materials, and composting would do a much better job. Stewart said, "I did this to illustrate an important point. The Mayor is not recycling his garbage at all, so he doesn't realize what a simple task it is."

July 30

Greenpeace Luxembourg protested outside a Du Pont chemical plant, which from 1988 intended to release high-level emissions of one of the chlorofluorocarbons, CFC 11, a chemical known to play a role in destroying the ozone layer. Large quantities of CFC 11 were to be used in the production of a new kind of plastic film.

August 4

Greenpeace demonstrators scaled the two 15-metre (50-foot) smokestacks of a garbage incinerator on the Tacoma Tideflats in Washington State,

August 4 Greenpeace climbers on the chimneys at Tacoma Tideflats garbage incinerator.

August 5 From the vantage point of the Trinity, Greenpeace and the Dean of York Minster survey the damage done to the cathedral by acid rain.

and plugged them with giant makeshift corks to dramatize their contribution to air pollution. Greenpeace also hung two banners: "Pollution Control Courtesy Of Greenpeace" and "Don't Use Our Lungs For A Landfill".

August 5

The Dean of York Minster cathedral went aloft in Greenpeace's hot-air balloon *Trinity* to see the damage that air pollution has done to the limestone fabric of the cathedral, which is in sight of three of the country's largest power stations. This event was part of a summer-long campaign aimed at documenting acid rain damage in the UK, during which the Greenpeace balloon flew over power stations and lakes, forests and historic buildings.

August 6

At midday today Dow Chemicals announced an end to their production of the herbicide 2,4,5-T, for which Greenpeace New Zealand had been campaigning for many years. Dow cited commercial reasons for the decision, saying that they considered it a safe and useful product, but that the costs of medical tests and the continued opposition of major environmental groups had seriously affected profitability. Production was to be halted by December 1987, and existing stocks were expected to be used up by late 1988.

August 19

Six Greenpeace activists chained themselves to a pier in Vancouver's inner harbour to prevent the docking of five nuclear-armed US warships. Police arrived with bolt cutters and all six protesters were arrested and charged with mischief.

August 28

Greenpeace activists blocked the waste water pipe of the BBU chemical company in Carinthia, Austria, close to the Italian border.

September

Philadelphia became the focus of Greenpeace's campaign against the export of toxic waste when two climbers scaled the 110-metre (360-foot) tower of the City Hall and hung a 150-square-metre (1,600-square-foot) banner that read: "Don't Export Toxic Ash to Panama. No Evenene Panama. (Don't Poison Panama)". The following day Greenpeace USA's toxics co-ordinator Pat Costner met with Philadelphia mayor Wilson Goode to present a report entitled "Burnt Offerings" – a Greenpeace analysis of the city's incinerator ash which, it claimed, was laden with heavy metals and dioxins.

The story had begun in 1986 when 10,000 tonnes of incinerator ash from the city had left on board the cargo ship *Khian Sea* on what was to become a year-long search for a dump site. After the cargo had been rejected by the Bahamas, Bermuda, Honduras, the Dominican Republic and Guinea-Bissau (in West Africa), the ship entered Haiti with a permit to import "fertilizer". The ash was partially unloaded before the authorities discovered its true nature

and ordered it to be removed. The ship left surreptitiously in the middle of the night, leaving at least 1,000 tonnes of the ash next to the waterfront.

At the time of the protest, Philadelphia officials were planning the first instalment of a 250,000-tonne shipment of ash to Panama, ash that had been refused a dump site in seven US states. As a result of Greenpeace lobbying, the Panamanian government also barred the shipments.

September 2

On the day that 21 nations took on a legal commitment to achieve a 30 per cent cut in emissions of sulphur dioxide by 1993, under the United

August 28 "*This poison pipe has been closed because of acute danger to the environment.*"

Nations Convention on Long Range Transboundary Air Pollution, Greenpeace wrote to the UK environment minister condemning the fact that Britain had refused to sign the protocol.

September 18

The European parliament passed a resolution calling for a ban on imports of all kangaroo products except for those from three species – the eastern grey, the western grey and the red – for which it called for better monitoring of, and documentation on, imports.

September 23

Greenpeace campaigners in Canada and the US released documents leaked to them by an official of the American Paper Institute, showing that the paper industry had known for more than a year that a variety of commonly-used paper products, including coffee-filters, disposable nappies, paper towels and tampons, are contaminated with small quantities of dioxin. The dioxin is formed in the chlorine bleaching process used in a certain type of paper mill, of which there are 90 throughout the States.

Greenpeace had previously released its report "No Margin of Safety" on August 20, which showed that the US Environmental Protection Agency (EPA) had known since 1980 that pulp mills were major dischargers of dioxin pollution. Instead of implementing regulations, they had entered into a secret pact with the industry to carry out a leisurely "joint study" of just five mills, selected by the industry itself. This agreement delayed a National Dioxin Study ordered by Congress in 1983. The study report was finally released on September 24, and it revealed that dioxin contamination had been discovered at more than 100 chemical companies, in fish from 100 sites and in municipal incinerators.

September 24

Greenpeace UK launched a two-month North Sea campaign to prevent pollution "physically and peacefully". The UK is the only country that allows sewage sludge to be dumped in the North Sea and, after 1989, will be alone in dumping industrial waste there.

October

Four Greenpeace activists and Earth First! co-founder Mike Roselle hung a banner on South Dakota's Mount Rushmore memorial to call attention to the problem of acid rain. Declaring that the protesters had "violated a shrine of democracy", the local Federal District court judge sentenced the team to 90 days in jail, two years' probation and a total of nearly $3,000 in fines – the toughest sentence ever meted out for climbing the monument. Because Mike Roselle also refused to agree to the terms of probation, he served an extra 80 days.

October 2

After nearly two years of arbitration, a panel of judges chaired by Claude Reymond, Professor of International Law at the University of Geneva in Switzerland, directed France to pay Greenpeace $5 million for the loss of the *Rainbow Warrior* and $1.2 million in "aggravated damages", as well as other costs and legal fees – a total of $8.16 million.

The case set a significant precedent in international law, representing the first time that a sovereign government had agreed to binding arbitration with a private, non-governmental organization over a dispute concerning actions taken by its military forces. This precedent, according to Greenpeace's lawyer Lloyd Cutler, "will hold state security and intelligence agencies to account in the future".

October 28

Greenpeace Netherlands, in conjunction with the Dutch broadcasting services, staged a second massive telethon (the first was in October 1985), comprising 10 hours of radio and television coverage as part of a campaign that raised more than $1 million.

Telethon II, focusing attention on the work of Greenpeace in the Netherlands, in the North Sea and worldwide, formed the finale to Greenpeace's national schools' campaign organized with the help of Dutch Educational Television. Some 50,000 schoolchildren turned their classrooms into Greenpeace campaign offices, participated in projects to solve local environmental problems, and supported the *Sirius*'s North Sea campaign through exhibitions on pollution and the work of Greenpeace.

By now Greenpeace Netherlands had become the biggest environmental organization in the country, with 18 employees, 300,000 supporters and a further 260,000 occasional donors.

November

Greenpeace USA, Alabamians for a Clean Environment and the Minority Peoples' Council organized a national protest against Alabama's Emelle landfill, probably the largest toxic waste landfill in the world. It is not only the dumping ground for toxic waste from 46 states, but it also receives waste from other countries through shipments from military bases. The demonstration coincided with the discovery that water from wells used to monitor chemical seepage was contaminated with toxic substances.

December 11 "Let the Whales Live" – a protest in Yokohama harbour.

GREENPEACE ITALY

The seventeenth Greenpeace national office in the world, in Italy, was opened in late 1986 and became operative the following year, beginning with a campaign against the nuclear plant at Latina and the transport of radioactive waste from there to Sellafield.

Later that year, in the final days of an electoral campaign for a referendum in which the Italian people were to be asked to support or oppose the use of nuclear power plants and the construction of new ones, Greenpeace activists succeeded in climbing the main chimney of the Latina plant and unfolding a banner protesting against the Italian Energy Department's involvement in the Superphenix fast breeder project.

Also in 1987, Greenpeace Italy exposed the existence of a trade in toxic waste leaving Italian harbours for disposal at various sites in the Third World.

This information campaign centred, in 1988, on the MV *Zenobia*, a vessel that had been travelling the world carrying chemical waste in search of a dump site. Such actions have gained Greenpeace more than 5,000 supporters in Italy.

November 14

East German police arrested five members of Greenpeace in Dresden after they unveiled a banner protesting against East German pollution of the River Elbe. The banner read: "Pollution Is Without Frontiers – Dresden – Hamburg – North Sea – Water Is Life."

November 30

Hans Guyt returned to Britain from the Netherlands and surrendered to the High Court. He began a three-month prison sentence for breaking

November 30 Hans Guyt – jailed for blocking the pipe at the Sellafield reprocessing plant.

December 13 The Rainbow Warrior *sinks to her final resting place in the waters of Matauri Bay.*

injunctions not to interfere with pipelines at the Sellafield plant.

December 11

Police in Yokohama, Japan, took six Greenpeace demonstrators and a 15-metre (50-foot) inflatable whale into custody. This followed an action next to the Japanese whaling ship *Nishin Maru 3* in protest at plans to send the boat to Antarctica for three months to kill 300 minke whales for "scientific" purposes.

December 13

The *Rainbow Warrior* was finally laid to rest about 24 metres (80 feet) down in Matauri Bay, on the eastern coast of New Zealand's North Island, at a ceremony attended by over 70 craft. The boat will form an artificial reef and provide a sanctuary for marine life. New Zealand's tourism minister, Phil Goff, said: "Today is not a time for bitterness but a time to pay tribute to the causes this ship has come to symbolize. The beauty and

peacefulness of this environment make it a fitting place to lay the *Rainbow Warrior* to rest."

A day later a furious diplomatic row broke out between France and New Zealand over the fact that Alain Mafart, one of the two French agents convicted of sabotaging the *Rainbow Warrior*, had been flown home from the Pacific atoll of Hao. The French claimed that he was seriously ill with kidney trouble.

Under the terms of the settlement reached between the two countries, neither of the two DGSE agents was to leave the island without the agreement of both governments.

BAN THE BURN

In August 1987, Greenpeace launched an eight-week action campaign against ocean incineration, the burning of hazardous waste, particularly highly toxic polychlorinated biphenyls (PCBs), in furnaces aboard ships at sea.

Greenpeace had been campaigning on the issue for five years in both Europe and the US, presenting national and international government forums with extensive scientific and technical evidence that calls into question the desirability, efficiency and environmental acceptability of waste incineration at sea.

During this period, Greenpeace succeeded in: getting one incinerator ship's licence revoked after discovering that it was illegally emitting dioxin into the atmosphere; preventing permits being issued for the incineration of PCB-contaminated waste in the Gulf of Mexico and the North Atlantic; defeating proposals for the construction of hazardous waste storage facilities at sites in the UK, the Netherlands and Belgium.

This latest Greenpeace initiative began with the release of a report detailing the potentially devastating effects of a large-scale spill of toxic waste at sea.

Boarding Party

Then, on August 21, activists from the *Sirius* boarded the incinerator ship *Vesta* in the North Sea, chaining themselves to the chimneys of the furnaces and hanging a banner. As a result, the ship was forced to take its full cargo of waste back to port.

Two days later, a late-night effort to board the ocean incineration ship *Vulcanus II* resulted in a nine-hour battle between activists in inflatables and the ship's crew armed with high-pressure hoses, who prevented Greenpeace from boarding.

The *Vulcanus II* subsequently headed for Spain, only to find that

pressure of public opinion, influenced by convincing evidence from Greenpeace, had led the Spanish government to revoke a permit allowing the ship to burn its cargo off the northern coast of Spain.

Forced to return to the North Sea, the *Vulcanus II* was once again prevented from burning its cargo, this time by the *Sirius* and a flotilla of some 30 Danish fishing boats. High-pressure hoses were deployed in a prolonged confrontation that ended on October 19, when the *Vulcanus II*'s propeller accidentally snagged a fishing net and the ship had to be towed back to harbour.

All this activity was designed to influence the deliberations on ocean incineration at the International North Sea Ministers' Conference, due to be held in London during

"Ban the Burn" Sirius *crew members hang a banner on the incinerator chimney of the* Vesta *(top). Actions by Greenpeace repeatedly prevented the* Vulcanus II *(above) from burning its cargo of hazardous waste at sea.*

November. At this meeting, the North Sea states agreed to reduce the amount of waste burned at sea by "not less than 65 per cent" by the end of 1990. This signalled the end of a profitable ocean incineration business in Europe. In January 1988, Waste Management Inc., owner of the *Vulcanus II*, announced that it was abandoning any plans to burn toxic waste in US waters.

The industry is now trying to re-establish the technology elsewhere, principally in Southeast Asia and the South Pacific.

GREENPEACE SPAIN

In 1987, following the successful Mediterranean campaign with the Sirius *the previous year, Greenpeace Spain hired a small sailing vessel, dubbed* Greenpeace V, *for a further series of actions.*

Leaving Barcelona on August 16, they first headed for the Tarragona petrochemical complex to highlight the pollution dangers it poses for the people living nearby. A similar purpose was served by sailing close to the buildings of the Vandellos nuclear reactors further down the coast.

The boat then visited Columbretes, where Greenpeace found the situation much improved from the previous year. The islands had been cleaned up, the replanting of native vegetation was under way and the Regional Parliament of Valencia was considering a proposal to preserve the area by declaring it a Maritime-Terrestrial National Park.

Hearing that military manoeuvres were to begin again in Cabrera, the boat headed there only to be ordered

"1992 Without Nukes" Spanish Greenpeace activists hang a banner bearing the title of their anti-nuclear report on the National Nuclear Security Council buildings.

by the Spanish Navy to return to Barcelona, where she was held in harbour. After a week, the crew decided to leave the vessel and mount an operation on land instead.

On June 5, Greenpeace volunteers wearing anti-pollution outfits and

gas masks dumped a tonne of toxic waste from the Peñarroya mining operation's discharge pipe at Portman Bay outside the Ministry of Public Works in Madrid.

In September, further actions took place against the dump ship *Nerva* and, in October, the Greenpeace ship *Moby Dick* sailed into the Mediterranean to take up the Cabrera issue once more. As a result, the military manoeuvres planned for the year were cancelled. The pressure increased when Greenpeace leaked a copy of the confidential scientific report commissioned by the Ministry of Defence, which confirmed that the manoeuvres were damaging the islands and recommended that the area be declared a National Park.

While the *Moby Dick* was in the area, Greenpeace took the opportunity to stage two actions in support of the nuclear-free seas campaign. On October 10, two inflatables attempted to stop the nuclear-armed destroyer USS *Comte de Grasse* berthing in Palma de Mallorca harbour. Later that month, the *Moby Dick*, together with a pontoon towed by an inflatable, blockaded the entrance of the United States naval base at Rota in southern Spain.

GREENPEACE ARGENTINA

The first Greenpeace office to be established in a developing country, Greenpeace Argentina has faced new challenges.

The main office in Buenos Aires was officially inaugurated on April 1, 1987, but work had begun as far back as February 1986 when Georgina Gentile and a group of volunteers began cutting through the red tape in order to register Greenpeace as a non-profit foundation.

The first National Board was created in November 1986 with Melvyn Gattinoni as president and Raúl Montenegro as vice-president.

In a country where a large sector of the population strives to make ends meet, it was important to define campaign priorities, to avoid being dismissed as too idealistic. Therefore the toxics problem was chosen for the main campaign, because it has a direct bearing on the quality of life of the average Argentinian.

The first step was to undertake a thorough research programme. The information gathered allowed Hugo Castello, the National Campaign Director, to set out the main goals: to achieve a ban on the production, importation, sale and use of the "Dirty Dozen" (12 chemicals used in pesticides identified as being damaging to health), and to promote legislation that would end dangerous and irresponsible waste-

disposal practices. The target areas were the Rio Negro Valley for pesticides and the Bahia Blanca Petrochemical complex for waste.

On another front, Raúl Montenegro, the Nuclear Campaign Director, established his goal: to block the building of the Gastre Nuclear Repository in Chubut Province, Patagonia. Greenpeace Argentina hopes to achieve this by lobbying for a legal ban on the importation of nuclear waste. Gastre is being built to handle not only Argentina's waste but that of other countries as well. Greenpeace is also working to alert the public to the dangers of a nuclear repository, to push for a referendum on the building of the plant, and to try to block international financial aid for the project.

A NEW TEAM TO ANTARCTICA

In the southern summer of 1987/1988, the MV Greenpeace *returned to Antarctica in order to resupply the "World Park" base on Ross Island. The winterers, who had spent almost 12 months in isolation, were brought on board, together with all the scientific material gathered during the year, ready for their return to New Zealand.*

Various modifications were made to the base complex, including the installation of solar and wind powered generators. A new "bio-loo" was set up to simplify the removal of all human waste, and improvements were made to the satellite communication system.

The new team of Greenpeace winterers were: leader, Keith Swenson, who had previously worked at the US McMurdo base as mechanic and survival instructor; Dr Wojtek Moskal, responsible for field research, with degrees in oceanography and meteorology; radio operator Sjoerd Jongens, who had spent considerable time in Antarctica working on upper-atmosphere physics with the Australian National Antarctic Research Expedition; and geologist Sabine Schmidt, who worked on various projects concerning coastal fish and plankton, pollution and other forms of human impact on the environment.

After returning to New Zealand, the MV *Greenpeace* travelled to Argentina and then to the Antarctic Peninsula. There, for one month, the crew visited many of the national bases in the area to check for environmental damage.

The MV *Greenpeace* hosted a meeting at which staff from the Chinese, Soviet and South Korean bases met with scientists from West Germany to consider the future of the continent. While moored near McMurdo base, the expedition ship was visited by over 100 US personnel who donated more than $1,000 to Greenpeace.

Radio Dome *A new tower improves satellite communications for the winterers at the Greenpeace base.*

Winterers *The new team are (left to right) Sjoerd Jongens, Keith Swenson, Wojtek Moskal and (foreground) Sabine Schmidt.*

1988

February

Greenpeace UK launched a billboard campaign and lobbied major supermarkets in an attempt to stop Icelandic whaling through a boycott of Icelandic fish products.

March 21

To mark the end of the European Year of the Environment, Greenpeace hung a banner on the Berlaymont in Brussels, opposite the council building where the EEC environment ministers meet. The banner, which carried a picture of a tree gradually losing its foliage, represented the death of the European forests.

March 30

Two Greenpeace climbers reached the top of the 52-metre (170-foot) Nelson's Column in London's Trafalgar Square to protest about the government's attitude to acid rain.

April

Admiral Pierre Thireaut, the French naval commander in the Pacific, announced plans to move part of the French nuclear test programme to Fangataufa, southwest of Moruroa. This revived suspicions that the nuclear blasts beneath Moruroa have created fissures in the atoll's coral reef and that radiation may be leaking into the Pacific.

April 23

Greenpeace Germany held an action at Plensburg on the German-Danish border against the nuclear-capable British frigate HMS *Brave*. Greenpeace personnel from six countries boarded the vessel, each wearing a T-shirt printed with one letter of the slogan "Atom Bombs On Board". The Royal Navy will neither confirm nor deny the presence of nuclear weapons on their ships.

April 25

Greenpeace released research linking the incineration of toxic waste at sea with increased levels of

poisons on the surface and seabed of the North Sea. The new evidence, which counters the waste-disposal industry's view that ocean incineration is clean and safe, was presented to scientists from the London Dumping Convention, the international body regulating the dumping of waste of all kinds.

On the same day, seven members of Greenpeace from Luxembourg, Belgium and Switzerland, dressed in radiation protection suits and masks, chained themselves to six yellow barrels, like those used to transport radioactive material, and blocked the entrance to Luxembourg near the city of Dudelange. Their protest, which took place on the second

March 21 "Stop Acid Rain" – a Greenpeace banner decorates the Brussels Berlaymont building.

anniversary of the nuclear accident at Chernobyl, was designed to draw attention to the dangers presented by the French nuclear power plant at nearby Cattenom.

April 29

In the UK, the Advertising Standards Authority rejected seven complaints by British Nuclear Fuels Ltd against advertisements carried in national newspapers about the imprisonment of the Greenpeace campaigner Hans Guyt for blocking the Sellafield waste pipeline.

GREENPEACE NORTH AMERICAN · INLAND WATERS EXPEDITION ·

GREAT LAKES · MISSISSIPPI

BELUGA

Water for Life

1988

May 5 "Water for Life" – a sticker marks the launch of the North American Inland Waters Expedition.

May 5

Greenpeace launched its North American Inland Waters Expedition, a six-month campaign jointly organized by national offices in Toronto and Washington.

Using the European riverboat *Beluga*, the expedition began in Montreal and then travelled up the St Lawrence and Saguenay rivers, spending two weeks sampling the water for toxic contaminants, which were analysed in the shipboard laboratory, and spotlighting the companies responsible.

The next two months were spent on the Great Lakes, highlighting the hazards posed by chemical discharges at 70 toxic "hotspots".

May 6

French Prime Minister Jacques Chirac staged an election campaign surprise by arranging for the second *Rainbow Warrior* saboteur, Captain Dominique Prieur, to be repatriated, without the consent of New Zealand's government or the UN, on the grounds that she was pregnant.

On the same day, Greenpeace activists operating from the *Moby Dick*, moored in international waters just outside the Baltic Sea, used inflatables to approach the Soviet warship *Silny* and attach a warning flag to her stern to signify she was carrying nuclear weapons.

May 26

At a press conference held on board the *Sirius*, Greenpeace demanded an end to the toxic discharges into the Mediterranean from the petro-chemical complex at Tarragona, Spain. For more than 20 years the chemical companies Bayer, BASF, Monsanto, Dow, Hoescht, Rio Tinto and others have been discharging harmful substances such as organo-chlorines and heavy metals directly into the sea from underwater pipes, in an area that has a developing tourist industry and is the base for one of the largest fishing fleets in the Mediterranean.

May 27

The *Moby Dick* sailed into Cork, in the Republic of Ireland, to lobby politicians on marine pollution. Principal Greenpeace concerns were radioactive pollution of the Irish Sea from Sellafield and the contamination caused by the dumping of industrial waste by the Pfizer chemical corporation off the Cork coast.

May 27 The Moby Dick sails to the Republic of Ireland to draw attention to pollution in the Irish Sea.

GREENPEACE

MOBY DICK

May 28

As part of Greenpeace's international campaign to end the mining and use of uranium, two Greenpeace activists – Michele Nanni and Dave Augeri – climbed to the top of the uranium conversion facility at the Eldorado nuclear production plant in Port Hope, Ontario, and hung a large banner reading: "Eldorado's Waste Is Everyone's Burden".

Eldorado has dumped almost a million tonnes of low-level radioactive waste at a dozen locations in the area. Port Hope's inner harbour now contains 90,000 cubic metres (120,000 cubic yards) of radioactive sediment, and has been designated an official "area of concern", that is, a toxic waste site requiring immediate attention to protect the Great Lakes from deadly pollution.

May 30

Four Greenpeace climbers abseiled down the front of the Auckland Sheraton hotel where the IWC meeting was taking place and unfurled a banner reading: "Stop Bloody 'Scientific' Whaling" in English and Japanese. Other protesters placed a giant inflatable whale and a huge silhouette of a harpooner on top of the building.

May 31

Further action took place against the dump ship MV *Kronos*, which was set to discharge its cargo of toxic waste in the North Sea. Despite being served with an injunction, Greenpeace pursued the ship with two inflatables as it left Nordenham, West Germany. One activist climbed the discharge pipe at the stern of the *Kronos* and managed to stay there for about 45 minutes.

On the same day, as part of the Mediterranean campaign (now in its third year), the crew of the *Sirius* finished taking samples from 10 underwater pipes belonging to various chemical companies that discharge waste into the waters of the bay of Tarragona.

This week, Greenpeace's Nuclear Free Future campaign focused

specifically on the British government's plans to build a second pressurized water reactor (PWR) at Hinkley Point in Somerset. The site already has two reactors – a Magnox and an advanced gas-cooled reactor (AGR).

Greenpeace began its campaign by publishing a report outlining the inadequacy of Central Electricity Generating Board (CEGB) emergency plans in the event of a major accident at Hinkley, and revealed details of how such a disaster would affect the local community. The report's release was timed to coincide with the start of a five-week tour by the *Moby Dick*.

The following day, Greenpeace lobbyists attended the pre-enquiry meeting into the Hinkley PWR and, on June 2, successfully executed a direct action in protest at the transportation of spent nuclear fuel from Hinkley to Sellafield. They unfurled a banner reading "Stop Hinkley-C. No More Waste" over a spent fuel container in a railway marshalling yard close to a primary school at Bridgwater in Somerset.

Six days later, the *Moby Dick* arrived at the island of Anglesey, off the coast of Wales, where Greenpeace held a public meeting to discuss CEGB plans to build another PWR at Wylfa on the island.

May 30 "Stop Bloody 'Scientific' Whaling" – a dramatic Greenpeace message to the IWC meeting.

June 2

Plans for a "minerals regime" for Antarctica were agreed, following a meeting of the Antarctic Treaty Consultative Parties in Auckland, New Zealand. This meeting was lobbied by Greenpeace, concerned that the regime opens the door to the full-scale exploitation of Antarctic mineral resources.

The new Convention on the Regulation of Antarctic Mineral Resource Activities permits prospecting for oil and minerals by seismic testing and other "low impact" techniques, and sets out rules that constitute a code for mineral exploitation. However, these rules appear to accept that there will be damage to the environment, and even seek to limit liability for such damage.

One ray of hope remains: the exploitation of mineral resources in Antarctica is unlikely to become economical in the near future, and Greenpeace may yet be able to persuade participating nations of the need for greater sensitivity to the ecological issues. The battle to preserve Antarctica continues.

ACTION LAB

On May 3, 1988, the Greenpeace action-bus took to the road in Denmark at the start of a 10-week tour of the seven countries that border the Baltic Sea, in a campaign aimed at encouraging an international approach to the area's environmental problems.

Eighteen metres (59 feet) long, the action-bus carries the most sophisticated mobile laboratory in Europe. Built and equipped by Greenpeace Austria to their own specifications, it first went on the road in July 1987.

The Baltic Sea has slow water circulation, low water temperatures throughout much of the year and no tides. As a result, it is a unique and particularly sensitive ecosystem, which is easily unbalanced by pollution of any kind.

The major sources of pollution are: waste discharges from the pulp and paper industry; sewage, which is a major cause of the shortage of oxygen in the water; and toxic waste from chemical industries, in particular sludge from the manufacture of titanium dioxide.

Over 140 million people inhabit the Baltic coast or live along the rivers flowing into it. Parts of the Baltic are already showing signs of environmental stress, with uncontrolled blooming of toxic algae and the presence of a virus that is affecting the Baltic seals, threatening their very survival.

During the 10-week tour, the Greenpeace campaigners used the equipment in the action-bus laboratory to take samples of polluted water and to document environmental abuse, as well as to attract the attention of both the public and the world's media to the issue.

One of the countries Greenpeace visited was the USSR, where the bus was the main attraction at a trade fair for marine technology in Leningrad. In Poland, Greenpeace worked with four Polish scientific institutions to analyse water samples from the Gdansk bay. This is one of the most grossly polluted areas in the Baltic and in Poland, and it is almost devoid of plant and animal life.

Baltic Bus *Greenpeace Austria has built the most sophisticated mobile laboratory in Europe.*

GREENPEACE IRELAND

Greenpeace Ireland was officially opened on May 12 following numerous visits by various Greenpeace boats over the years, and the establishment of a temporary office there the previous year.

Sellafield is the major environmental issue in Ireland. Since the blocking of the plant's discharge pipe on June 3, 1987, Greenpeace has concentrated on developing a political approach for its campaign. The first stage was to attempt to have the Irish government take legal action against the United Kingdom, and a brief outlining a number of legal options was distributed to political contacts. The options included actions within the British legal system, within the European Court, and within the International Court of Justice at the Hague.

In October 1987, to mark the thirtieth anniversary of the Windscale (Sellafield) fire, representatives from all Irish political parties attended a Greenpeace press conference to demand an end to reprocessing at Sellafield; they also took part in an event organized by CORE (Cumbrians Opposed to a Radioactive Environment).

Many Irish local councils are now actively involved in the campaign to shut Sellafield, as are many schools. In November 1987, 150 pupils from a Dublin school travelled to England to hand in a petition at 10 Downing Street, calling for Sellafield's closure.

The Irish office is also involved in: the Nuclear Free Seas Campaign; attempts to get Ireland to change its position on ocean incineration (it presently supports this method of waste disposal); protesting against the annual cull of grey seals by fishermen; and examining the problems facing dolphins in Irish waters.

June 4

Greenpeace was one of 94 organizations and individuals in 59 countries to receive awards for "outstanding environmental achievement" from the United Nations Environmental Programme (UNEP). Other recipients included tree planters in rural Asia, scientists in the Middle East, wildlife experts in Africa, environmental campaigners in Latin America and conservation lobbyists in North America. UNEP intends to make further awards in order to compile a "Global 500" roll of honour by 1991.

June 9

The *Sirius* and four of her inflatables intercepted the nuclear aircraft carrier USS *Dwight D. Eisenhower*, part of the Sixth Fleet, on her arrival at Palma de Mallorca in the Balearic Islands in the Mediterranean. The *Eisenhower*, which is nine times as long as the *Sirius*, carries 100 nuclear bombs for her aircraft and 25–30 nuclear depth charges. The ship is powered by two atomic reactors.

The following day, the *Sirius* was arrested in Ibiza harbour, and a military guard was placed on board. She was released two days later.

June 13

A new Greenpeace boat, the *Rubicon*, set off from Brighton on the south coast of England on a tour of Cornish, Welsh, Irish and Scottish waters, as part of the UK Dolphin Lookout Scheme. The voyage was designed to monitor and film the behaviour of dolphins in coastal waters, and document the various environmental hazards that pose a growing threat to the existence of these small cetaceans.

The *Rubicon*, an 11-metre (36-foot) steel-hulled, gaff-rigged cutter, was purchased in 1987 for £10,000 by Greenpeace UK.

June 9 The gigantic nuclear aircraft carrier USS Dwight D. Eisenhower *dwarfs a Greenpeace inflatable.*

June 17

Greenpeace activists in Spain, Denmark, Sweden and Belgium carried out a coordinated protest against plans to burn 2,000 tonnes of Spanish toxic waste on the ocean incineration vessel *Vulcanus II* in the North Sea. In Spain, two Greenpeace volunteers, together with members of a local environmental group, unfolded a banner over one of the toxic waste containers due to be loaded aboard the *Vulcanus* in Santander. In Denmark, four people chained themselves to the Spanish Embassy doors, and there were other similar actions in Sweden and Belgium.

Four days later, two crew members from the *Moby Dick*, which had been diverted from her planned

FLEET FOR THE FUTURE

In 1988, Greenpeace welcomed two new ships to the fleet, one for "World Park" Antarctica and another for the warmer waters of the Pacific Ocean.

The 499-tonne, 51-metre (167-foot) MV *Viking*, a West German-built supply vessel, has a double-skinned hull, making her suitable for use in Antarctic waters. Greenpeace has renamed her *Gondwana*, after the supercontinent to which Antarctica once belonged. The ship's refit included the construction of a helicopter landing pad and hangar, and the addition of a superstructure to provide extra accommodation, a communications centre and a reception area. A hold was created to carry supplies for the Antarctic "World Park" base. The initial cost of the vessel was $1.2 million, and the conversion work cost a further $1 million.

The *Grampian Fame*, a 30-year-old British fishing vessel, has become the new *Rainbow Warrior*. With the fitting of new efficient engines and a modern sailing rig, she provides Greenpeace with a ship capable of operating in the Pacific for extended periods. In 1989, after an initial tour of Europe, the *Grampian Fame* will

Gondwana *A new Greenpeace vessel is renamed for Antarctica.*

sail to the Pacific, to campaign against French nuclear testing, driftnet fishing, the trade in toxic waste, missile flight testing over the Pacific and the destruction of delicate coral reef ecosystems.

June 17 Spanish Greenpeace activists hang a banner at a depot that stores toxic waste to be burned aboard the Vulcanus II.

activities in the Irish Sea, met with violence when they attempted to board the *Vulcanus II* in the North Sea, 160 kilometres (100 miles) off Scarborough. One of them was struck with a pair of wire-cutters and the other was blasted into the sea by high-powered water jets. The jets also damaged the bridge of the *Moby Dick*, destroying electrical and radar equipment.

June 21

Greenpeace activists in Helsinki chained themselves for eight hours to eight containers of Icelandic whale meat being sent illegally to Japan, and called on the Finnish authorities to seize the shipment and destroy it. The containers, which were awaiting transport by rail through the Soviet Union to Japan, held 196 tonnes of meat from endangered fin and sei whales. The authorities later agreed to send the consignment back to Iceland.

Also on this day, in protest at the US Congress's slow progress on enacting legislation to help prevent acid rain, four Greenpeace climbers scaled two smokestacks at a coal-fired power plant at Chalkpoint, Maryland, which supplies power to Washington DC.

June 24

The *Sirius* entered the US submarine base at La Maddalena, off the northeast coast of Sardinia, and launched two inflatables that headed for a submarine tender that carries and services US Tomahawk cruise missiles. Despite attempts to stop them, the Greenpeace activists managed to tie a "yellow submarine" to the anchor-chain of the tender to symbolize non-violent use of the seas. The submarine, made of nylon and timber, carried two crew members – Heather Holve of the US and Guilia Fusco of Italy – who stayed with their vessel even when it was rammed by a patrol boat crewed by sailors carrying boat-hooks, grapples and knives.

June 27

At an international conference in Toronto, Canada, on "The Changing Atmosphere – Implications for Global Security", Greenpeace announced its intention to launch an international campaign in 1989 to protect the atmosphere, centring initially on halting the depletion of the ozone layer.

June 29

Two Greenpeace climbers scaled the Department of National Defence building in Ottawa to protest at the Canadian government's planned purchase of nuclear submarines. They hung banners from the building that read: "Nuclear Subs – Deep Trouble – Greenpeace". The purchase of the submarines would cost $30 billion.

July 2

By diving from inflatable dinghies and creating a human blockade in the water, Greenpeace activists delayed the entry of the 4,500-tonne nuclear-capable USS *Conyngham* into the Danish port of Aalborg. For eight hours the activists swam in front of the destroyer and hung on its anchor lines while the *Conyngham*'s crew blasted them with high-powered water hoses. The captain of the *Conyngham* had refused to declare whether or not his ship was carrying nuclear weapons, a requirement under Danish law. Twenty-five people were arrested and charged with disturbing shipping traffic and breaking harbour regulations.

July 11

Seven *Sirius* crew members blocked the pipe and channel used by the Belgian multinational company Solvay to discharge industrial waste into the Mediterranean from its factory at Rosignano in northwest Italy. The activists, wearing gas masks and anti-pollution suits, were ferried ashore in inflatables. Chaining themselves to the sides of the channel, they hung a banner and plugged the waste discharge pipe with a bung. All seven campaigners were later arrested.

July 12

The *Moby Dick* began a tour of nuclear naval bases around the British coast as part of Greenpeace's Nuclear Free Seas campaign. The

July 11 "*Stop the Dumping of Poison in the Sea*" – *the Solvay protest.*

launch coincided with the publication of a Greenpeace report that called for a radical reassessment of emergency procedures in the event of a nuclear accident at the Rosyth naval base in Scotland, a dockyard that refuels and maintains nuclear-powered submarines and is the home port for four Type 42 destroyers capable of carrying nuclear depth charges.

July 29

Four climbers (Renato Ruf of Switzerland, Torben Holm Lauridson of Denmark, Holger Spiegel of Germany and Luc Stoffel of Luxembourg) scaled the chimneys at the Arbed-Belval steel production plant and hung a 25-metre (82-foot) banner reading "Stop Air Pollution". The plant is the largest emitter of sulphur dioxide in Luxembourg, and a major contributor to the problem of acid rain.

Three days later, Luc Stoffel and Paul Lepesant, of Greenpeace Luxembourg, climbed Arbed's administrative building and hung another banner: "The Forest Cries -

Arbed Remains Silent". Five more protesters bearing placards blocked an entrance to the building and, with the intercession of the mayor of Luxembourg City, persuaded Arbed to hold discussions with Greenpeace on air pollution and acid rain.

August 10

While a Greenpeace inflatable distracted a police patrol boat, the *Vega* pulled alongside the USS *Roarke* in Vancouver harbour, Canada, and Greenpeace activist Simon Waters quickly boarded the warship to determine whether she was carrying nuclear weapons in the nuclear weapon free zone of Vancouver City. Waters was taken into custody before he could complete his inspection.

August 11

Greenpeace Italy launched "Project Buccaneer" against illegal trawl fishing in shallow coastal waters, which threatens the breeding grounds of many species, by placing artificial reefs off the harbour of Civittavecchia. (A similar action took place two weeks later, at the island of Elbe, northeast of Corsica.) The action was supported by the

August 21 The Rubicon *sails the British coast on dolphin patrol.*

Department for the Marine Environment, the Ministry of the Merchant Navy, the harbour authorities, and an environmental management company, which supplied a vessel to put the reefs in position. Within three weeks, two illegal trawlers were arrested and their equipment confiscated.

On the same day, following the death of thousands of common seals in the North Sea and the Baltic, Greenpeace held an emergency conference in London, at which European scientists examined the possible causes of the killer disease. Baltic and North Sea waters are known to be contaminated with a complex mixture of toxic chemicals, some of which are discharged in millions of tonnes annually. The scientists cited evidence from studies of other mammals showing that polychlorinated biphenyls, dioxin and other persistent organic compounds accumulate in seals and can reduce immunity to disease.

Greenpeace issued a statement of concern, calling for an immediate end to the dumping of radioactive and toxic wastes in these waters and announced an "Emergency Push" campaign to achieve this.

August 21

The *Rubicon*, on her dolphin-watch tour of British coastal waters, put into Cromarty on the east coast of Scotland, where the crew spent a week holding meetings with local dignitaries and members of the press. Greenpeace proposed the setting up of a locally-organized dolphin watch scheme, and the establishment of protected areas for small cetaceans in the Firth.

August 26

Climbers from Greenpeace Norway scaled the 160-metre (525-foot) chimney of Borregaard Industries' pulp and paper factory in southern Norway to hang a banner saying: "Nice Promises – Dead Seas". As well as discharging heavy metals and other toxic waste directly into the Glomma River, and thence into the Kattegat area of the North Sea, Borregaard is also the area's biggest single source of atmospheric pollution, emitting some 175 tonnes of sulphur dioxide every year.

Also on this day, as part of Greenpeace's "Emergency Push" on pollution in the Baltic and North Sea, members of Greenpeace Germany dumped 20 tonnes of blinded fish, covered with sores, at the headquarters of the Finnish Kemira group of companies in Helsinki. The Kemira group, which is 99 per cent government-owned, is the world's fifth largest producer of titanium dioxide.

The fish were caught near the Vuorikemia works on the west coast of Finland, which discharges 700 tonnes of sulphuric acid, iron sulphate and titanium dioxide into the North Sea every day.

September 9

Two Greenpeace members took the issue of the seal plague and North Sea pollution to the British prime minister. Wearing gas masks and protective clothing, they draped a dead seal with a banner saying "North Sea R.I.P" and carried it to Downing Street. They were protesting at government inaction during the North Sea epidemic that had already claimed 12,000 seals on the continental coastline and 600 in the UK. A seal in the Irish Sea had also died from the same virus.

October 4

Four members of Greenpeace in Luxembourg, in response to Japan's

September 9 *Greenpeace delivers a dead seal to Downing Street.*

plan to kill 300 minke whales "for research" in the Antarctic, suspended Flo the inflatable whale from a conference building in which the Japanese ambassador was giving a lecture. Flo, illuminated by spotlights, bore a banner reading: "Japan - Stop Killing Whales".

Meanwhile in Antarctica, the Greenpeace team mounted a protest against pollution by the US McMurdo base. A large quantity of garbage had been blown the 19 kilometres (12 miles) across the ice of McMurdo Sound, and had ended up plastered against the walls of the Greenpeace base and entangled in their equipment. Keith Svenson interrupted a meeting between National Science Foundation officials and new arrivals at the US base to return the rubbish and ask that steps be taken to prevent pollution in the area. The other Greenpeace activists hung a banner saying: "Trash On The Ice – Keep McMurdo In One Place".

October 6

Dutch and German Greenpeace members drove four inflatables around the bow of the Swedish nuclear transport vessel *Sigyn* to prevent her docking at Emden, West Germany. Despite radio calls from Greenpeace and the firing of distress flares, the *Sigyn* proceeded into port, crushing one of the inflatables under her hull. The two occupants just managed to jump clear.

Later in the day, Greenpeace had cause to celebrate when delegates from over 65 countries, at a meeting of the London Dumping Convention in England, agreed to halt all ocean incineration of toxic waste by the end of 1994.

1989

March

This month saw the launch of "Greenpeace: Breakthrough", a double album featuring hit tracks donated by 24 leading rock musicians and bands. With an initial run of 3 million copies, it was not only the biggest release of a Western rock record in the USSR, but was also the first record to be issued simultaneously in the Soviet Union and the rest of the world.

Among the artists included are: Peter Gabriel, Bruce Hornsby and the Range, Sade, the Pretenders, Dire Straits, Eurythmics, John Cougar Mellencamp, Bryan Ferry, Simple Minds, Talking Heads, Belinda Carlisle, the Waterboys, Aswad, the Thompson Twins, Grateful Dead, Sting, Terence Trent D'Arby and Bryan Adams.

Proceeds from the album are to be used to establish an office for Greenpeace in the Soviet Union and to promote cooperative environmental projects between the eastern and western blocs.

In the words of David McTaggart, who delivered the master copy of the record and the cover artwork to Moscow, "This album is a great big greetings card to young people in the Soviet Union from Greenpeace and the artists.

"It's going to introduce them not only to the best of Western rock but also to the idea of people working together across national boundaries to build a safe, clean world."

March *David McTaggart with Bryan Adams, one of the 24 artists featured on the "Greenpeace : Breakthrough" album being released in the USSR.*

INDEX

Page numbers in **bold** refer to illustrations and captions

BIBLIOGRAPHY

We acknowledge our debt to the following books on the subject of the history of Greenpeace:

Greenpeace, Robert Keziere and Robert Hunter. McClelland and Stewart Ltd, Toronto 1972.
Greenpeace III, Journey Into The Bomb, David McTaggart with Robert Hunter. Collins, London 1978.
To Save A Whale, The Voyages of Greenpeace, Robert Hunter and Rex Weyler. William Heinemann, London 1978.
The Greenpeace Book, Karl and Dona Sturmanis. Orca Sound Publications, Vancouver 1978.
The Greenpeace Chronicle, Robert Hunter. Picador, London 1980.
A Ma Mer. Greenpeace/Casterman, Paris 1984.
Greenpeace In Actia, Anne Boermans. Meulenhoff Informatief, Amsterdam 1985.
The Rainbow Warrior Collection, Ponga Tree Press, New Zealand 1986.
Eyes of Fire, The Last Voyage of the Rainbow Warrior, David Robie. Lindon Publishing, Auckland 1986.
Sink The Rainbow, John Dyson. Victor Gollancz Ltd, London 1986.
The Death of the Rainbow Warrior, Michael King. Penguin, London 1986.
Rainbow Warrior, The Sunday Times Insight Team. Arrow Books, London 1986.
The Rainbow Warrior Affair, Richard Sears and Isobelle Gidley. Unwin Paperbacks, London 1986.
The Greenpeace Book of Antarctica, John May. Dorling Kindersley Ltd, London 1988
Greenpeace, Erik Claudi. Tiderne Skifter, Copenhagen 1986.

GREENPEACE OFFICES

GREENPEACE ARGENTINA
Junin 45, 3 Piso
1026 BUENOS AIRES
ARGENTINA

GREENPEACE AUSTRALIA
Postal:
Private Bag 6, Post Office
Broadway
Sydney, NSW 2007
AUSTRALIA

Street:
4/134 Pioneer House
Broadway
Sydney, NSW 2007
AUSTRALIA

Regional Offices:
155 Pirie Street
Adelaide SA 5000
AUSTRALIA

C/-102 Bathurst Street
Hobart
TASMANIA 7000

GREENPEACE AUSTRIA
Mariahilfer Gurtel 32
1060 Vienna
AUSTRIA

GREENPEACE BELGIUM
Waversesteenweg 335
1040 Brussels
BELGIUM

GREENPEACE CANADA
578 Bloor Street West
Toronto, Ontario M6G 1K1
CANADA

Regional Office:
2623 West 4th Avenue
Vancouver BC V6K 1P8
CANADA

Regional Office:
Street:
5611 Clark
Montreal PQ, H2T 2V5
CANADA

Postal:
CP 472
Succursale-E
Montreal PQ, H2T 3A9
CANADA

GREENPEACE DENMARK
Head Office:
Thomas Laubs Gade 11-13
2100 Kobenhavn O
DENMARK

Information Offices:
Greenpeace-Kobenhavn
1358 Kobenhavn O
DENMARK

Greenpeace-Aarhus
Guldsmedgade 20
8000 Aarhus C
DENMARK

GREENPEACE GERMANY
Vorsetzen 53
D-2000 Hamburg 11
WEST GERMANY

GREENPEACE IRELAND
29 Lower Baggot Street
Dublin 2
EIRE

GREENPEACE ITALY
28 viale Manlio Gelsomini
00153 Rome
ITALY

GREENPEACE JAPAN
Contact Office:
501 Shinwa Building
9-17 Sakuragaoka
Shibuya-ku
Tokyo 150
JAPAN

GREENPEACE LATIN AMERICA
Apartado 230
Centro Colon
San Jose 1007
Costa Rica
CENTRAL AMERICA

GREENPEACE LUXEMBOURG
Street:
24 Rue Dicks
L-4081 Esch/Alzette
LUXEMBOURG

Postal:
Postbox 229
L-4003 Esch/Alzette
LUXEMBOURG

GREENPEACE NETHERLANDS
Damrak 83
1012 LN Amsterdam
NETHERLANDS

GREENPEACE NEW ZEALAND
Private Bag
Wellesley Street
Auckland
NEW ZEALAND

GREENPEACE NORWAY
St Olavsgt 11
PO Box 6803 St Olavsplass
0130 Oslo 1
NORWAY

GREENPEACE SPAIN
C/Rodriguez San Pedro 58, 4 Piso
28015 Madrid
SPAIN

Regional Office
Passeig Maritim 44
07015 Palma de Mallorca
SPAIN

GREENPEACE SWEDEN
PO Box 7183
S-402 34 Goteborg
SWEDEN

Regional Office:
Box 7629
S-103 94 Stockholm
SWEDEN

GREENPEACE SWITZERLAND
Muellerstrasse 37
Postfach 4927
8022 Zurich
SWITZERLAND

GREENPEACE UK
30-31 Islington Green
London N1 8XE
ENGLAND

GREENPEACE USA
1436 U Street, NW
Washington DC, 20009
USA

Great Lakes:
1017 W.Jackson Boulevard
Chicago, Illinois 60607
USA

Northeast:
139 Main Street
Cambridge, MA 02142
USA

Southeast:
Suite 80
Wilton Plaza
1881 NE 26th Street
Wilton Manors, FL 33305
USA

Northwest:
The Good Shepherd Center
4649 Sunnyside Avenue,
N Seattle, WA 98103
USA

Pacific Southwest:
Fort Mason, Building E
San Francisco, A 94123
USA

Photographic credits

The following organizations and individuals provided photographs and gave permission for them to be reproduced.
Associated Press Ltd. (**68** bottom, **113** side, **125** top); *Auckland Star* (**17**, **18**, **115**, **118** top and bottom); *Gert Eggenberger/Pressefoto* (**141**); *Al Giddings/Ocean Images, Inc.* (**32**, **42/43**); *Gil Hanly* (**119**); *NASA/ Science Photo Library* (**128**); *New Zealand Herald* (**24**, **26** top); *Vancouver City Archives* (**8**); *Vancouver Sun* (**6**)

GREENPEACE: *Kurt Abrahamson* (**99** bottom); *Doug Allen* (**131** bottom right); *Douglas Baglin* (**87**); *Peter Ballem* (**50** bot); *Fernando Baptista* (**123** bottom); *Erol Baykal* (**48**, **50** top); *Michael Chechik* (**79**); *David Cross* (**132**); *Richard Dawson* (**91** middle, **92** top, and bottom); *Deloffre* (**60** bottom); *Koen Dom* (**105** bottom right); *Tom Donaghue* (**99** side); *Jean-Paul Ferrero* (**56** bottom); FRI CREW (**30**); *Jacqueline Geering* (**142** bottom); *Pierre Gleizes* (**4** right, **55** bottom, **61**, **62**, **63**, **69** side, **70** middle, **71** bottom, **73** bottom, **74** top and bottom, **75** top left, top right and bottom left, **76** top, **77** top and side, **81**, **82**, **83** bottom, **84**, **86**, **88** top and side, **93**, **94** top and side, **96** top, **97** top, **103** top and middle, **106** top); *John Goldblatt* (**155** top); *Miguel Gremo* (**151**, **153** top); *Robert Hines* (**130** top); *Kenn Hollis* (**106** side); *Ann-Marie Horne* (**28/29** sequence); *Kematen* (**133** top); *Andrew Kerr* (**140** top); *Robert Keziere* (**3** left, **10**, **11** top and bottom, **12**, **13** top, **15**); *Kiner* (**5** left); *Therese Kristensen* (**100**); *Pieter Lagendijk* (**70** top, **89** bottom); *Brian Latham* (**143** top); *Lawrence* (**130** bottom); *David McTaggart* (**22**, **78**); *Maso* (**113** bottom); *Miller* (**116/117**); *Patrick Moore* (**45** bottom right, **46** bottom, **52** main); *Steve Morgan* (**112** side, **148** bottom); *Myers* (**5** right); *Lasse Spang Olsen* (**104** bottom right); *Paczensky* (**98** side); *Klaus Pahlich* (**105** top and bottom left); *John Parulis* (**90**, **91** top); *Ron Pemberton* (**138** bottom); *Fernando Pereira* (**102**, **107**, **108** bottom, **109** top and bottom, **110/ 111** all pictures); *James Perez* (**137**, **140** bottom left, **146** top and bottom); *Tim Peters* (**97** side); *Campbell Plowden* (**56** top, **101** bottom); *Dick Powys* (**53**); *Ros Reeve* (**154**); *Ribaut* (**104** top); *David Rinehart* (**45** top, **80**); *Lisa Schaublin* (**139**); *Sandy Scheltema* (**129**); *Seiffe* (**104** bottom left); *Solara* (**101** side); *Leslie Stone* (**149**); *Kazumi Tanaka* (**59** bottom); *Jaap van der Veer* (**144** middle, **152** top); *Diether Venneman* (**4** left, **136**); *Tomas Walsh* (**138** top); *Westerling* (**147**); *Rex Weyler* (**34**, **35**, **36** top and bottom, **37**, **38** top and bottom, **39** top, **40** top, **42** top, **45** bottom left, **51** side, **54** top, **58**, **65** bottom, **76** bottom); *Zeiller* (**150** bottom). All other photographs are by Greenpeace photographers whose names we have been unable to ascertain, and to whom we apologize.

The following artists have given permission for their illustrations to be used: *Jean "Moebius" Giraud* (**51**); *Alain Goutal* (**64**); *Ferdinand Guiraud* (**124** - originally appeared in *Canard Enchaîné,* November 6, 1985); *Randolph Holmes* (**41**); *Mike* (**49** side - sticker based on his artwork); *Terry Peters* (**65** top side); *Roy Peterson* (**31**); *Roland Sabatier* (**25** bottom); *William Stout* (**44** - © William Stout 1988, used with permission of *Last Gasp*, San Francisco).

The maps were drawn by *Janos Marfy*
Picture research by *John May*

Strenuous efforts have been made by the authors and editors to ensure that proper accreditation has been given and that copyright clearance has been obtained on all material in this book. We will happily amend any omissions for future editions.

Acknowledgements

JOHN MAY Any history is the product of many hands, this one more than most. It has been a long and difficult project, the making of which is a story in itself.

Thanks first of all to my co-author Michael Brown, who travelled around the world interviewing key campaigners and assembling the mass of information that became the first draft of this book. His valuable work made this book possible.

The manuscript then went through four more complete drafts as we struggled to contain a monstrous amount of fresh information that seemed to pour in from all quarters. Here tribute should be paid to the work of Lesley Riley and Ian Whitelaw, who worked tirelessly and way beyond the call of duty to maintain control and fashion order out of chaos.

Numerous people within Greenpeace gave willingly of their time and memories on this project. Principally we wish to thank David McTaggart, Steve Sawyer, John Frizell and Jim Bohlen.

The photo-librarians at Greenpeace - Jay Townsend in the US and Jacqueline Geering and Sarah Saunders in the UK- supported us unfailingly in this project. We would have been lost without their efficient and professional response to our constant requests.

The following Greenpeace people also deserve thanks for their support, encouragement and information:

Peter Bahouth, Denise Bell, John Bowler, Sandra van den Brink, Leslie Busby, Andre Carothers, Maureen Coles, Duncan Currie, Dick Dillman, Lorrette Dorreboom, Cornelia Durrant, Peter Dykstra, Florian Faber, Maureen Falloon, Brian Fitzgerald, Nancy Foote, Nick Gallie, Melvyn Gattinoni, Georgina Gentile, Martin Gotje, Monika Griefahn, Charlotte Grimshaw, Hans Guyt, Henk Haazen, Bertil Hagarhall, Lena Hagelin, Sebia Hawkins, Jon Hinck, Ron van Huizen, Bruce Jaildagian, Helen Kingham, Athel von Koettliz, Renata Krosa, Jakob Lagercrantz, Elaine Lawrence, Gerd Leipold, Jane McAslan, Joyce McClean, Dan McDermott, Bunny McDiarmid, Michael Manolson, Tony Marriner, Elke Martin, Micki Mathias, Julie Miles, Doug Mulhall, Susi Newborn, Michael Nielsen, Goeran Olenborg, Marc Pallemaerts, Philip Parker, Remi Parmentier, Xavier Pastor, Teresa Perez, Alan Pickaver, Beverley Pinnegar, Kelly Rigg, Elaine Shaw, Michelle Sheather, Jonathan Smales, Roger Spantz, Gianni Squitieri, Mia Stenstrom, Carol Stewart, Lorraine Thorne, Allan Thornton, Cookie Timmel, Louis Trussell, Peter Whitehouse, Peter Wilkinson, Neville Williams, Wladimir Zalozieckyj, Filip van Zandijcke, Harald Zindler.

Special thanks to Martin Leeburn and Sean Mac Nialluis at Greenpeace Communications for making our operation possible. Thanks also to Raewyn McKenzie who did such a sterling job of research for us in New Zealand, and to Jean-Marc Piaf in Paris.

This note would not be complete without thanks to all our friends in Lewes, who have helped lighten our burdens, and especially to Tanya, Alex and Louis.

MICHAEL BROWN extends thanks to Ann-Marie Horne, Robert Hunter and Patrick Moore for their help, and to all at Greenpeace offices in Vancouver, Washington, Lewes and New Zealand.

DORLING KINDERSLEY would like to thank Peter Brookesmith for his generous help with the text, Sean Moore for his work on the proofs and Hilary Bird for compiling the index. Jane Warring deserves special acknowledgement for her skilful design of the book and the commitment she showed to it.

While the book was in production we were saddened to learn of the death of John Cormack, the first Greenpeace skipper, on November 17, 1988. He was born in Vancouver, British Columbia, on October 24, 1912.

WE SEE THIS BOOK as the beginning of the process of documenting Greenpeace's complex and dynamic history. The story is still unfolding. In the absence of organized archives and with key memoirs yet unwritten, this is unlikely to be the last word.

We welcome your comments, corrections, additions and amendments, which will help us to make subsequent editions of this book even more comprehensive.

This book is dedicated to the hundreds and thousands of people who have made Greenpeace a force to be reckoned with and to the millions that support their work worldwide. May it add strength to your efforts and deepen your convictions.